To Dr. Robert W. Hobson II
with our compliments and thanks
for continued leadership and
collaboration in this important
research field of ischemia-reperfusion.

AF880596

Walter Durán

Peter J. Pappas M.D.

Medical Intelligence Unit

Ischemia-Reperfusion Injury of Skeletal Muscle

Representative photomicrograph of porcine rectus abdominis muscle (400X). At baseline, the venular lumen is filled with erythrocytes and neutrophils do not adhere to the endothelium. Leukocytes are absent from the interstitium. (Fig. 3.1a, from Kirschner RE and Fantini GA, Chapter 3, Role of Neutrophils *© 1994.)*

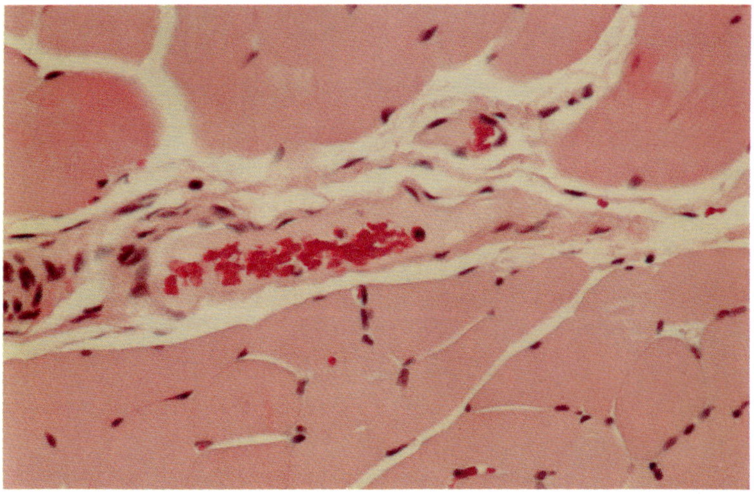

Following six hours of ischemia and one hour of reperfusion, numerous neutrophils are seen adherent to the venular endothelium. (Fig. 3.1b, from Kirschner RE and Fantini GA, Chapter 3, Role of Neutrophils *© 1994.)*

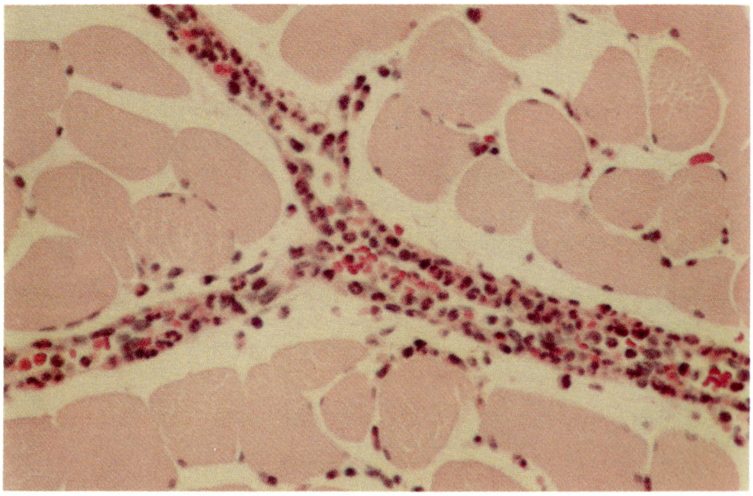

At four hours of reperfusion, there is extensive migration of neutrophils from the venular lumen into the interstitium. (Fig. 3.1c, from Kirschner RE and Fantini GA, Chapter 3, Role of Neutrophils *© 1994.)*

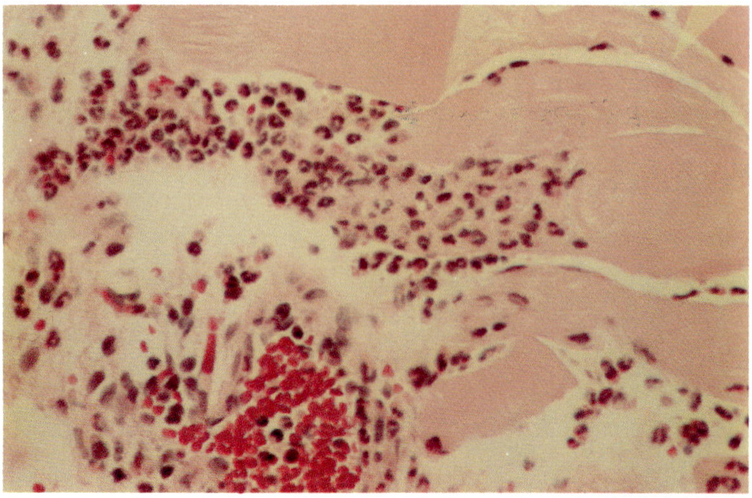

MEDICAL
INTELLIGENCE
UNIT

ISCHEMIA-REPERFUSION INJURY OF SKELETAL MUSCLE

Gary A. Fantini, M.D.

Cornell University Medical College
New York, New York

R.G. LANDES COMPANY
AUSTIN

Medical Intelligence Unit

ISCHEMIA-REPERFUSION INJURY OF SKELETAL MUSCLE

R.G. LANDES COMPANY
Austin

CRC Press is the exclusive worldwide distributor of publications of the Medical Intelligence Unit.
CRC Press, 2000 Corporate Blvd., NW, Boca Raton, FL 33431. Phone: 407/994-0555.

Submitted: March 1994
Published: June 1994

Production Manager: Judith Kemper
Copy Editor: Constance Kerkaporta

Printed in the U.S.A.

Please address all inquiries to the Publisher:
R.G. Landes Company, 909 Pine Street, Georgetown, TX 78626
or
P.O. Box 4858, Austin, TX 78765
Phone: 512/ 863 7762; FAX: 512/ 863 0081

ISBN 1-57059-131-8
CATALOG # LN9131

Library of Congress Cataloging-in-Publication Data
Ischemia-reperfusion injury of skeletal muscle / [edited by] Gary A. Fantini.
p. cm.—(Medical intelligence unit)
Includes bibliographical references and index.
ISBN 1-57059-131-8(hard cover)
1. Reperfusion injury. 2. Ischemia. 3. Muscles—Pathophysiology.
4. Muscles—Blood-vessels. I. Fantini, Gary A. II. Series.
[DNLM: 1. Muscles—blood supply. 2 Muscles—physiopathology. 3. Reperfusion Injury.
4. Ischemia. WE 500 I775 1994]

RB144.5.I83 1994
611'.0186—dc20
DNLM/DLC
for Library of Congress

94-15167
CIP

CONTENTS

1. **Historical Perspective** **1**
Malcolm O. Perry

2. **Reactive Oxygen Metabolites and their Consequences** **5**
Tadafumi Akimitsu, Dean C. Gute, Sarah N. Jerome and Ronald J. Korthuis
   1. The Concept of Reperfusion Injury 6
   2. Biochemistry of Reactive Oxygen Metabolites 6
   3. Role of Reactive Oxygen Metabolites in the Genesis of Postischemic Skeletal Muscle Injury 9
   4. Neutrophils Contribute to the Pathogenesis of Muscle I/R..... 13
   5. Reactive Oxygen Metabolites Promote Postischemic Leukocyte Infiltration 18
   6. Mechanisms of Oxidant-Mediated Cellular Injury 19
   7. Postischemic Capillary No-Reflow 20

3. **Role of Neutrophils** **33**
Richard E. Kirschner and Gary A. Fantini
   1. Neutrophil Infiltration 33
   2. The Role of Neutrophil Adhesion 34
   3. Mechanisms of Neutrophil-Mediated Injury 42
   4. The No-Reflow Phenomenon 49
   5. Conclusions 52

4. **Platelet Activating Factor** **61**
Peter J. Pappas and Walter N. Durán
   1. History 62
   2. Biosynthesis 62
   3. Mechanisms of Signal Transduction 64
   4. Second Messenger Production Mediated by Phospholipid Turnover 64
   5. Protein Kinase C and Tyrosine Kinase 65
   6. Summary of Signal Transdcution 65
   7. Cellular Responses and Effects of Cytokines 66
   8. Microcirculatory Effects of PAF 67
   9. Effects of PAF during Ischemia/Reperfusion 72

CONTENTS

5. **Complement Activation** 77
Thomas F. Lindsay, Barry B. Rubin, Alexander D. Romaschin and Paul M. Walker
1. Complement 77
2. The Classical Pathway 78
3. The Alternate Pathway 79
4. Membrane Attack Complex 80
5. Regulation of the Complement Cascade 81
6. Complement Activation Following Tissue Ischemia 81
7. Complement Activation and Skeletal Muscle 82
8. Activation of the Alternate Complement Pathway Following Experimental Muscle Ischemia 83
9. Complement Blockade Reduces Skeletal Muscle Permeability after Ischemia 87
10. Summary 91

6. **Tumor Necrosis Factor and Interleukin-1** 93
Mark Gennaro and Enrico Ascer
1. Basic Information 94
2. Cytokines and Leukocytes 94
3. Cytokines and Endothelial Cells 97
4. Summary 99

7. **Nitric Oxide (Endothelium-Derived Relaxing Factor)** 103
Edith Tzeng and Timothy R. Billiar
1. Biosynthesis and Mechanism of Action 105
2. Nitric Oxide Synthases 106
3. NO as the Endothelium-Derived Relaxing Factor 108
4. NO and Free Radicals 110

8. **Iron Delocalization and Iron Binding** 125
Bo E. Hedlund
1. Iron Delocalization and Reoxygenation Injury 125
2. Therapeutic Intervention for Attenuation of Iron Mediated Injury 134

9. **Remote Effects: Lung Injury** 143
John M. Davis and Gary A. Fantini
1. Historical Perspective 144
2. The Complement System 144
3. Role of the Neutrophil 144
4. Leukotrienes 146
5. Pulmonary Effects of Remote I/R: Clinical Investigation 148
6. Summary 149

Index 157

CONTRIBUTORS

Editor

Gary A. Fantini, M.D., Assistant Professor of Surgery
Cornell University Medical College
Assistant Attending Surgeon (Vascular)
The New York Hospital, New York, New York
Chapter 3, 9

Tadafumi Akimitsu, M.D.
Department of Physiology and Biophysics
Louisiana University Medical Center
School of Medicine in Shreveport
Shreveport, Louisiana
Chapter 2

Enrico Ascer, M.D.
Vascular Surgeon
Maimonides Medical Center
Brooklyn, New York
Chapter 6

Timothy R. Billiar, M.D.
Department of Surgery
University of Pittsburgh
Pittsburgh, Pennsylvania
Chapter 7

John Mihran Davis, M.D.
Associate Professor of Surgery
Cornell University Medical College
Associate Attending Surgeon
The New York Hospital
New York, New York
Chapter 9

Walter N. Durán, Ph.D.
Professor of Physiology and Surgery
Chief, Division of Microcirculatory Research
UMDNJ-New Jersey Medical School
Newark, New Jersey
Chapter 4

Mark Gennaro, M.D.
Vascular Surgeon
Maimonides Medical Center
Brooklyn, New York
Chapter 6

Dean C. Gute, D.V.M., Ph.D.
Department of Physiology and Biophysics
Louisiana University Medical Center
School of Medicine in Shreveport
Shreveport, Louisiana
Chapter 2

Bo E. Hedlund, Ph.D.
Biomedical Frontiers, Inc.
Minneapolis, Minnesota
Chapter 8

CONTRIBUTORS

Sarah N. Jerome, B.S.
Department of Physiology
and Biophysics
Louisiana University Medical Center
School of Medicine in Shreveport
Shreveport, Louisiana
Chapter 2

Richard E. Kirschner, M.D.
Department of Surgery
Cornell University Medical Cellege
New York, New York
Chapter 3

Ronald J. Korthuis, Ph.D.
Department of Physiology
and Biophysics
Louisiana University Medical Center
School of Medicine in Shreveport
Shreveport, Louisiana
Chapter 2

Thomas F. Lindsay, M.D., M.Sc.
Division of Vascular Surgery
The Toronto Hospital
Department of Surgery
University of Toronto
Toronto, Ontario, Canada
Chapter 5

Peter J. Pappas, M.D.
Vascular Fellow
UMDNJ-New Jersey Medical School
Newark, New Jersey
Chapter 4

Malcolm O. Perry, M.D.
Professor of Surgery
Chief, Division of Vascular Surgery
Texas Tech University
Health Sciences Center
Lubbock, Texas
Chapter 1

Alexander D. Romaschin, Ph.D.
Division of Vascular Surgery
The Toronto Hospital
Department of Surgery
University of Toronto
Toronto, Ontario, Canada
Chapter 5

Barry B. Rubin, M.D., Ph.D.
Division of Vascular Surgery
The Toronto Hospital
Department of Surgery
University of Toronto
Toronto, Ontario, Canada
Chapter 5

Edith Tzeng, M.D.
Department of Surgery
University of Pittsburgh
Pittsburgh, Pennsylvania
Chapter 7

Paul M. Walker, M.D., Ph.D.
Division of Vascular Surgery
The Toronto Hospital
Department of Surgery
University of Toronto
Toronto, Ontario, Canada
Chapter 5

CHAPTER 1

Historical Perspective

Malcolm O. Perry

In the book entitled "The Lives of a Cell," Lewis Thomas made the following observation regarding scientific progress: "Great new blocks of information are being brought in almost daily and put precisely down in what were previously empty spaces".[1] During the past two decades, ischemia-reperfusion injury of skeletal muscle has come under intense scrutiny, and new information collected from clinical and laboratory investigation has shed light on the pathophysiology of this problem. It was previously believed, and often repeated, that resting skeletal muscle had a low requirement for energy and oxygen as compared to more active organs which were busy with their assigned tasks, and therefore, skeletal muscle could tolerate up to six hours of total ischemia without permanent damage.[2] This myth of a "safe interval" of ischemia has been discredited. Observations from the experimental laboratory and operations in humans have clearly shown that even temporary periods of partial ischemia can result in damage to the cell, causing injuries which may be difficult to detect when viewing the overall function of the organ.[3-5] Even without the risk of scattered intravascular thrombosis, which may complicate temporary ischemia, low flow states can result in muscle cell injury that often goes undetected. Eklof and his associates observed that temporary occlusion of the abdominal aorta during vascular operations resulted in disturbances in the energy systems of the muscles of the legs which persisted for at least sixteen hours.[6] Subsequent laboratory studies have confirmed muscle damage following short periods of partial as well as total ischemia. As these experiments progressed, it became clear that much of the injury was occurring after the onset of reperfusion, rather than during the ischemic interval.[7]

These observations can be correlated with histologic studies of ischemic muscle. Numerous laboratory experiments and a few clinical observations have shown that there are few histologic changes in muscle tissue during the period of ischemia; however such changes are manifest shortly after blood

Ischemia-Reperfusion Injury of Skeletal Muscle, edited by Gary A. Fantini, M.D.;

flow is restored.[8] These findings suggest that ischemia and reperfusion might best be viewed as two separate entities, each of which can damage muscle cells. The difference between ischemia and hypoxia is also clearly an indication of a separate pathophysiology. Cardiac muscle retains contractility even when subjected to rather severe hypoxia, however ischemia precipitates immediate cessation of contraction.

Autoregulation has been defined as the ability of an organ system to maintain constant blood flow despite wide variations in blood pressure. The kidney, for example, maintains a relatively constant renal blood flow over a range of blood pressures from 60 to 180 mm Hg. In contrast, however, skeletal muscle does not have this capacity to effectively maintain blood flow during periods of low perfusion pressure. Bayliss, in 1902, suggested that autoregulation was due to an intrinsic property of the smooth muscle cells in the vessel wall.[9] This myogenic theory has been accepted as the most likely explanation for this remarkable ability of certain organs to maintain adequate blood flow. There are factors which can modulate this response, including circulating catecholamines, sympathetic and parasympathetic stimulation, and perhaps other mediators as well. When blood pressure drops below 30mm Hg, Burton has shown that there is a tendency for small blood vessels to close.[10] Thus, there is a critical blood pressure for effective perfusion. Blood is a non-Newtonian fluid with an anomalous viscosity. This has been described by some as a "plastic" characteristic. Because of these hemodynamic features, a finite increase in blood pressure is required to restart blood flow from a state of resting inertia. This pressure increase can be higher then that which might be needed to maintain flow once initiated. Thus, it can be seen that after an interval of very low blood pressure, there are impediments to restoration of flow even in the absence of thrombosis.

In 1948, Harman studied the problem of reperfusion in skeletal muscle by injecting bromophenol blue into muscles which had been ischemic for varying intervals.[11] He examined the rate of penetration and elimination of the vital dye and found that it immediately stained normal and mildly ischemic muscles as soon as the tourniquet was released, indicating establishment of reflow. If the ischemic interval was longer than three hours, the dye was retained in the muscle. Following six hours of ischemia, the dye did not enter the muscle for at least 30 minutes. He concluded that there was stagnation of the circulation, and confirmed this by histologic evaluation, where he observed dilated capillaries engorged with red blood cells, but not fibrin and clots. This concept of a "no reflow" phenomenon was explored further by Ames, Flores and others who found that in the kidney and brain, as well as in other organs, tissue perfusion did not always return to normal, despite the restoration of normal blood pressure.[12,13] Histopathological examination of these tissues revealed endothelial blebs, swelling of surrounding supporting cells and structures, and an increase in edema.[14] Thrombosis did not appear to be the problem.

Subsequently, Ernst and associates investigated the possibility that the neutrophil might be involved in this "no reflow" phenomenon.[15] Observations of the pulmonary circulation in experimental hemorrhagic shock and in other low flow states revealed sequestration of white blood cells in the small vessels and plugging of capillaries. The white blood cell measures approximately 12μ in diameter and is relatively difficult to deform under the best of circumstances. However, in these pathologic situations, rheology may be altered and the mechanical effects of obstruction become more apparent. The relationship between the endothelial cell and the white blood cell, the importance of rolling and adhesion, and subsequent activation and release of mediators was not fully appreciated at this time.

Zimmerman, in a study of cardiac ischemia and reperfusion, observed that the cardiac muscle remained relatively intact if the perfusate did not contain calcium.[16] When calcium was reintroduced into the perfusate, marked damage to the cardiac muscle cells was noted. This was called the "calcium paradox". In these early stages of investigation, the role of calcium as a direct toxin and its pivotal role in proteolysis were not fully understood. What was clear from these experiments was that a noxious agent was involved. The problem of mediator-induced damage was emerging as an important part of systemic illness.

Early studies by McCord, Bulkley, Del Maestro, Granger, Parks, and their associates clearly established the role of oxygen-free radicals in these disease processes.[17-21] Bulkley pointed out that because oxygen is so ubiquitous in our environment, its toxic properties have not been fully appreciated.[22] The NADPH-dependent oxidase system on the surface of neutrophils is recognized as an efficient source of superoxide radicals. The "respiratory burst," a mechanism by which phagocytes destroy bacteria, is an oxygen-dependent process, although the bacteriocidal activity may be mediated through the secondary generation of hypochlorous acids. Free radicals in biology have been recognized for many years, but their endogenous production and capacity to injure were not apparent before these investigators published their work. Studies of ischemic intestine documented the danger from these toxic species, and subsequently Granger et al identified xanthine oxidase as the source of these radicals.[20] These experiments were expanded to include other organ systems including heart, liver, kidneys and skeletal muscle. Early studies suggested that xanthine dehydrogenase in skeletal muscle, while measurable, did not convert to the active form of xanthine oxidase upon nonperfusion. It was initially believed that this correlated with resistance of skeletal muscle to ischemic injury as compared to other tissues. It has subsequently been shown that the neutrophil and the endothelial cell contain metabolic processes which are capable of generating toxic oxygen free radicals,[22,23] and that the neutrophil itself is an important mediator of oxidative injury in postischemic skeletal muscle.[24,25] In fact, it is the interaction between the neutrophil and endothelial cell that has come to occupy center stage. These cells are found throughout the body, and no organ is protected from injury by aberrations in these oxygen systems.

Hechtman and his associates have shown, for example, that following lower limb ischemia there is sequestration and activation of neutrophils within the lungs and the limbs, but not in other organs.[26] In a recent article entitled "The Horror Autotoxicus and Multiple-Organ Failure," Baue describes the far-ranging effects of these toxic mediators and efforts to control this form of "biologic self-destruction".[27] There are, of course, many other mediators incriminated in these disease processes, including more than 30 different cytokines. Although tumor necrosis factor and the interleukins have received the most attention, it is clear that with time other mediators will be identified as major players in these morbid events.

There are now wide-spread efforts to understand and interrupt the interaction between the neutrophil and the endothelial cell. The adhesion of the white blood cell, its activation, emigration and eventual release of cytotoxic granules, are clearly of great importance in these processes. Receptor blockade with analogues, monoclonal antibodies, or simple filtration and depletion of neutrophils are techniques being explored to interrupt these sequences. There is an attempt to understand the relationship of the arachidonic acid pathway to the generation of oxygen free radicals, and to elucidate the roles that complement and platelet activating factor play in these pathologic events.

As the "previously empty spaces" receive their new blocks of information, the number of questions increases exponentially. Molecular oxygen and its relationship to life is at the very center of this investigation. Lewis Thomas pointed out that although mitochrondria and chloroplasts have remained small and conservative, "these two organelles are, in a fundamental sense, the most important living things on earth. Between them they produce the oxygen and arrange for its use. In effect, they run the place".[1]

References

1. Thomas L. The Lives of a Cell. New York, Viking Press, 1974:75-80.
2. Miller HH, Welch CS. Quantitative studies on the time factor in arterial injuries. Ann Surg 1949; 130:428-438.
3. Haljamae H, Enge E. Human skeletal muscle energy metabolism during and after complete tournequet ischemia. Ann Surg 1975:183-916.
4. Perry MO, Fantini G. Ischemia: Profile of an enemy. Reperfusion injury of skeletal muscle. J Vasc Surg 1987; 6:231-234.
5. Roberts JP, Perry MO, Hariri RJ et al. Incomplete recovery of muscle cell function following partial but not complete ischemia. Circ Shock 1985; 17:253-258.
6. Eklof B, Neglan P, Thompson D. Temporary incomplete ischemia of the legs induced by aortic clamping in man. Ann Surg 1980; 93:89-98.
7. Perry MO, Shires III GT, Albert SA. Cellular changes with graded limb ischemia and reperfusion. J Vasc Surg 1984; 1: 536-540.
8. Malan E, Tatoni G. Physio- and Anatopathology of acute ischemia of the extremities. J Cardiovasc Surg 1963; 4:2-10.
9. Bayliss W. On the local reaction of the arterial wall to changes of internal pressure. J Physiol 1902; 28:220-229.
10. Burton AC. On the physical equilibrium of small blood vessels. Am J Physiol 1951; 164:319-330.
11. Harman JW. The significance of local vascular phenomena in the production of ischemic necrosis in skeletal muscle. Am J Pathol 1949; 24:625-641.
12. Ames A, Wright RL, Kowada M et al. Cerebral ischemia–the no-reflow phenomenon. Am J Pathol 1968; 52:437-450.
13. Flores J, Di Bona DR, Beck CH et al. The role of cell swelling in ischemic renal damage and the protective effect of hypertonic solution. J Clin Invest 1972; 51:118-127.
14. Jameson RL. The role of cellular swelling in the pathogenesis of organ ischemia. West J Med 1974; 3:205-216.
15. Ernst E, Hammerschmidt DE, Bagg E et al. Leukocytes and the risk of ischemic diseases. JAMA 1987; 257:2318-2324.
16. Zimmerman ANE, Daenes W, Hahman WC et al. Morphological changes of heart muscle caused by successive perfusion of calcium-free or calcium- containing solutions (Calcium Paradox). Cardiovasc Res 1967; 1:201-209.
17. McCord JM. Oxygen-derived free radicals in postischemic tissue injury. N Engl J Med 1985; 312:159-163.
18. Bulkley GB. The role of oxygen free radicals in human disease processes. Surgery 1982; 94:407-411.
19. Del Maestro RF. An approach to free radicals in medicine and biology. Acta Physiol Scand [Suppl] 1980; 492:153-168.
20. Granger DN, Rutili G, McCord JM. Superoxide radicals in feline intestinal ischemia. Gastroenterology 1981; 81:22-29.
21. Parks DA, Bulkley GB, Granger DN. Role of oxygen free radicals in shock, ischemia, and organ preservation. Surgery 1983; 94:428-432.
22. Reilly PM, Schiller HJ, Bulkley GB. Pharmacologic approach to tissue injury mediated by free radicals and other reactive oxygen metabolites. Am J Surg 1991; 161:488-503.
23. Granger DN. Role of xanthine oxidase and granulocytes in ischemia-reperfusion injury. Am J Physiol 1988; 255:H1269-H1275.
24. Korthuis RJ, Grisham MB, Granger DN. Leukocyte depletion attenuates vascular injury in postischemic skeletal muscle. Am J Physiol 1988; 254:H823-H827.
25. Yokota J, Minei JP, Fantini GA et al. Role of leukocytes in reperfusion injury of skeletal muscle after partial ischemia. Am J Physiol 1989; 257:H1068-H1075.
26. Welbourn CRB, Goldman G, Paterson IS et al. Neutrophil elastase and oxygen radicals: synergism in lung injury after hind limb ischemia. Am J Physiol 1991; 260: H1852-H1856.
27. Baue AE. The horror autotoxicus and multiple-organ failure. Arch Surg 1992; 127:1451-1462.

CHAPTER 2

Reactive Oxygen Metabolites and their Consequences

Tadafumi Akimitsu

Sarah N. Jerome

Dean C. Gute

Ronald J. Korthuis

Prolonged ischemia results in skeletal muscle dysfunction and eventually tissue necrosis through loss of cellular energy charge and accumulation of potentially toxic products of cellular metabolism. Although reestablishing adequate blood flow is essential to salvage ischemic skeletal muscle, reperfusion is not without peril. Indeed, a growing body of evidence indicates that highly reactive metabolites of molecular oxygen are formed during reperfusion.[1-5] These reactive oxygen species are cytotoxic by virtue of their ability to induce alterations in the structure and function of essentially every class of biomolecule found in cells. Thus, oxidant formation during reperfusion may exacerbate tissue injury induced by ischemia. In addition, reperfusion induces oxidant-mediated, neutrophil-dependent microvascular barrier disruption, edema formation and the genesis of capillary no-reflow.[1,3-5] The increased nutrient diffusion distance induced by the development of these microcirculatory disturbances impairs cellular nutrition during reperfusion despite successful resolution of the event that initiated the reduction in blood flow in the first place.

Recognition of the fact that reperfusion may initiate pathologic reactions that increase the injury sustained by ischemic tissue has important clinical implications because it is the reperfusion component of tissue injury that is amenable to therapeutic intervention. Elucidation of the mechanisms involved in the production of postischemic tissue injury may provide important new avenues for the treatment of ischemia/reperfusion (I/R) injury. The purposes of this chapter are to describe the basic biochemistry of reactive oxygen metabolites and to summarize the evidence that we and others

Ischemia-Reperfusion Injury of Skeletal Muscle, edited by Gary A. Fantini, M.D.;

have accumulated in support of the hypothesis that reactive oxygen metabolites play an important role in the genesis of tissue injury in postischemic skeletal muscle.

THE CONCEPT OF REPERFUSION INJURY

Minimizing ischemic time by early restitution of blood flow has traditionally been viewed as the only important intervention for diminishing the extent of skeletal muscle injury induced by decreased muscle perfusion. However, the results of several recent studies indicate that reperfusion may paradoxically increase the extent of skeletal muscle injury and necrosis. For example, only minor histologic evidence of skeletal muscle injury is evident during ischemia or brief periods of reperfusion. In contrast, progressive injury occurs with longer periods of reperfusion.[6] In addition, therapeutic interventions that are initiated at the onset of reperfusion are as effective in reducing postischemic tissue injury as those introduced prior to ischemia.[7]

While the aforementioned studies clearly establish that reperfusion injury is distinct from that which occurs as a result of ischemia per se, the mechanisms involved in the production of postischemic tissue injury remain uncertain. However, a growing body of evidence indicates that skeletal muscle must be exposed to molecular oxygen during reperfusion to demonstrate the microvascular dysfunction and myocyte necrosis that occurs when blood flow is reinstituted. For example, we have demonstrated that reperfusion with normoxic blood induces microvascular barrier disruption and the development of capillary no-reflow.[8] In contrast, reperfusion with anoxic blood produces very little damage. Furthermore, several groups have reported that limited reintroduction of molecular oxygen during the first hour of reperfusion substantially reduces the extent of postischemic muscle necrosis.[9,10] These observations indicate that the pathologic events initiated by reperfusion require the reintroduction of molecular oxygen to produce tissue injury. It now appears that the cytotoxic events initiated at the onset of reperfusion involve the formation of highly reactive metabolites of molecular oxygen.

BIOCHEMISTRY OF REACTIVE OXYGEN METABOLITES

Before describing the evidence in favor of the view that reactive oxygen metabolites contribute to postischemic muscle injury, we will review the basic biochemistry of oxidants.[11,12] The term reactive oxygen metabolites (ROM, or species, ROS) has been introduced to describe any compound derived from molecular oxygen that has acquired less than four electrons. For example, the superoxide anion radical, hydrogen peroxide, and the hydroxyl radical are considered ROM because they have been reduced by one, two and three electrons, respectively. The complete four electron reduction of oxygen results in the production of water, which is not considered a ROM. Although often used synonymously with ROM, an oxygen-derived free radical is defined as a molecule or atom that contains one or more unpaired electrons in its outer orbital. Using this definition, superoxide is a free radical while hydrogen peroxide is not. Nevertheless, both are ROM. Thus, oxygen-derived free radicals are a subset of ROM.

Since the thermodynamic potential of unpaired electrons to form an electron pair is very high, ROM tend to be highly reactive (although there is a considerable variation in their reactivity). Most ROMs are potent oxidizing agents in that they tend to remove electrons from biomolecules. Superoxide is a bit peculiar in this regard in that it can also act as a reductant, donating its odd electron to transition metals or ascorbate which, in turn, may promote the formation of other more cytotoxic radical species.

As noted above, the addition of one electron to molecular oxygen results in the formation of the superoxide anion radical. Although superoxide can spontaneously dismutate to form hydrogen peroxide, the enzyme superoxide dismutase (SOD) greatly accelerates the rate ($\sim 10^4$ times) of this reaction.[13] While hydroxyl radicals can be formed via the interaction between superoxide and hydrogen peroxide, the rate of this reaction is too slow to be of biologic consequence. However, the rate of this reaction can be substantially enhanced by the presence of trace amounts of transition metals, e.g., iron, or metal chelates, e.g., ADP-Fe^{3+}, which act as catalysts.[11,12,14] This is the so-called iron-catalyzed Haber-Weiss or Fenton reaction. The hydroxyl radical is extremely reactive and interacts with the first molecule it encounters after its formation.[11,12,15,16] Therefore, the molecular alterations induced by hydroxyl radicals occur at its site of formation.

Alternative mechanisms for the formation of hydroxyl radicals that are iron-independent have also been proposed. Beckman et al[17] have suggested that the free radical nitric oxide (NO, an endothelium-derived relaxing factor) can interact with superoxide to form peroxynitrite which, at physiologic pH, decomposes to yield hydroxyl radicals and nitrogen dioxide. Ramos et al[18] have suggested that neutrophil-derived hypochlorous acid can interact with superoxide to form hydroxyl radicals by a mechanism that is iron-independent. However, no evidence has been presented to support the notion that either of these reactions occur in vivo.

Hemeproteins (myoglobin and hemoglobin) can interact with small amounts of hydrogen peroxide or lipid hydroperoxides to yield ferryl myo- or hemoglobin.[12] In the presence of larger amounts of peroxides, heme iron is released which may then redox cycle to perpetuate oxidant reactions. Although hemeprotein-associated free radicals may play an important role in the pathogenesis of muscle I/R, this possibility has not received a great deal of attention. However, it has been suggested that rhabdomyolysis may contribute to renal failure induced by reperfusion of ischemic skeletal muscle by a mechanism that involves the generation of a hemeprotein-associated oxidant.[19]

Sources of Reactive Oxygen Metabolites

Within reperfused postischemic tissues, there are multiple potential biologic sources of cytotoxic oxygen metabolites, including the enzyme xanthine oxidase, activated leukocytes, the mitochondrial electron transport chain, arachidonic acid metabolism, and catecholamine autoxidation. Of these, two enzymatic sources, xanthine oxidase and neutrophilic nicotinamide adenine dinucleotide phosphate (NADPH) oxidase, have received the most attention with regard to muscle I/R.

Xanthine Oxidase

The enzyme xanthine oxidase (XO) is widely distributed among mammalian tissues and participates in the oxidation of a wide variety of endogenous and exogenous substrates. This enzyme is best known for its roles in the terminal oxidation of purines and the absorption of iron by the gastrointestinal tract.[20] In addition, it has been suggested that XO may be a primary source of ROM in reperfused tissue.[20-25]

XO exists usually in nonischemic, healthy cells predominantly as an oxidized nicotinamide-adenine dinucleotide (NAD^+)-dependent dehydrogenase (XDH).[20,22,26] This form of the enzyme uses NAD^+ instead of molecular oxygen as the electron acceptor during oxidation of purines and does not produce superoxide or hydrogen peroxide.[20] However, XDH is converted to the oxidant-producing XO form (D-to-O conversion) during tissue ischemia in many tissues.[20,27,28] Additionally, ischemia induces the breakdown of adenosine triphosphate (ATP) to hypoxanthine and xanthine, resulting in a reduction in cellular energy charge and increased tissue levels of hypoxanthine and xanthine, both of which can act as substrates for the XO reaction.[29-33] Indeed, studies in skeletal muscle have demonstrated that ischemia results in a marked depletion of ATP and a several-fold increase in tissue hypoxanthine levels.[30,31,33] Ischemia-induced D-to-O conversion and the accumulation of hypoxanthine and xanthine set the stage for oxidant production by the XO reaction during reperfusion. Reperfusion supplies the second required substrate for XO, molecular oxygen. As a consequence, when molecular oxygen is reintroduced during reperfusion, ischemia-induced XO generates superoxide and hydrogen peroxide during the oxidation of hypoxanthine and/or xanthine that accumulated during the ischemic period.

D-to-O conversion can occur via two mechanisms: XDH is reversibly converted to XO by oxidation of thiol groups, or irreversibly converted through proteolysis of a small portion of the enzyme by a calcium-dependent protease which is activated by increased intracellular calcium that occurs secondary to the changes in metabolism during ischemia.[27,34,35] The notion that D-to-O conversion may occur by an irreversible, proteolytic mechanism is supported by the finding that this conversion can be inhibited by the protease inhibitor, soybean trypsin inhibitor.[25,36]

Ischemia-induced D-to-O conversion occurs at different rates in different organs. While relatively short periods of ischemia induce conversion in the intestine and liver, much longer periods are required in skeletal muscle.[20,22,37] For example, D-to-O conversion occurs within minutes of induction of ischemia in splanchnic organs but does not appear to occur in rat skeletal muscle subjected to less than one hour of ischemia. However, two hours of ischemia is associated with marked increases in muscle XO activity.[22,37] It should also be noted that the degree of D-to-O conversion is proportional to the duration of ischemia; i.e., longer periods of ischemia lead to larger increments in XO activity.[20,22,37]

Histochemical localization studies indicate that XO is localized in the sarcolemma and mitochondria of aerobic muscle fibers,[38] while immunolocalization techniques using antibodies directed against XO demonstrate that this enzyme is concentrated in large quantities in capillary endothelial cells in bovine and human skeletal muscle.[39,40] In fact, capillary endothelial cells contain more than 100 times the XO activity of other cells in the same tissue. Moreover, XO activity detected in cultured endothelial cells subjected to anoxia and reoxygenation is sufficient to produce microvascular injury, suggesting that XO may be a major source of cytotoxic reactive species in endothelial cells.[41]

The localization of XO to microvascular endothelial cells in skeletal muscle has important implications for reperfusion injury in this tissue. This localization implies that capillary endothelial cells should be especially susceptible to oxidant-mediated reperfusion injury,[40] and that this oxidant-producing enzyme is located at the same site that will be first exposed to oxygen (one of the substrates for XO) upon reperfusion.[22] Damage to the capillary endothelial cells may act as a stimulus to attract and activate inflammatory cells which cause further tissue damage.[42] XO activity of skeletal muscle is low compared with other tissues such as liver and intestine when measured in whole organ homogenates.[20,22,37] However, the measured levels of the enzyme are sufficient to cause endothelial cell injury.[43] Furthermore, the concentration of XO in microvascular endothelium might allow this enzyme to play a role in the pathogenesis of oxidant-mediated microvascular injury in postischemic skeletal muscle.

NADPH Oxidase

The concept that reperfusion promotes the recruitment of leukocytes into postischemic tissues is well established. Phagocytic leukocytes contain a membrane-bound nicotinamide adenine dinucleotide phosphate (NADPH) oxidase. This enzyme oxidizes cytoplasmic NADPH to $NADP^+$ and reduces molecular oxygen to superoxide, which can subsequently dismutate to form hydrogen peroxide.[14,44] Increased oxygen metabolism of activated neutrophils is known as the "respiratory burst".

Activated neutrophils also secrete a variety of granular enzymes, some of which are involved in production of oxidants. Myeloperoxidase (MPO) is one of these granular enzymes and catalyzes the reaction of hydrogen peroxide with chloride ions to form hypochlorous acid (HOCl). The hydrogen peroxide that fuels this reaction is derived from the spontaneous dismutation of neutrophil-derived superoxide. Although neutrophil-derived superoxide and hydrogen peroxide can generate hydroxyl radicals in the presence of trace amounts of transition metals, granulocytes have been shown to produce very few hydroxyl radicals in vivo.[45] The lack of hydroxyl radical production by activated neutrophils appears to be related to the fact that MPO consumes virtually all of the hydrogen peroxide thereby limiting the interaction of hydrogen peroxide with superoxide.

ROLE OF REACTIVE OXYGEN METABOLITES IN THE GENESIS OF POSTISCHEMIC SKELETAL MUSCLE INJURY

PROTECTION AFFORDED BY LIMITING THE AVAILABILITY OF MOLECULAR OXYGEN

The notion that ROM may contribute to the genesis of postischemic tissue injury was first proposed on the basis that reperfusion-induced injury can be reduced by limiting the availability of molecular oxygen, which is the source of these oxidants, when blood flow is reestablished. Indeed, reperfusion with normoxic blood causes microvascular injury in skeletal muscle whereas anoxic reperfusion produces very little damage (Fig. 2.1).[8] However, replacement of anoxic blood with oxygenated blood during reperfusion causes tissue injury similar to that seen with normoxic reperfusion alone. Similarly, reperfusion using an artificial plasma to which oxygen and erythrocytes are gradually reintroduced or limiting bloodflow during reperfusion to preischemic rates are effective in attenuating postischemic skeletal muscle necrosis (Fig. 2.2).[9,10,46] Analogous findings have been reported in other tissues such as heart and stomach.[9,47-49] The results of these studies confirm the concept that a substantial proportion of I/R injury occurs during the reperfusion phase

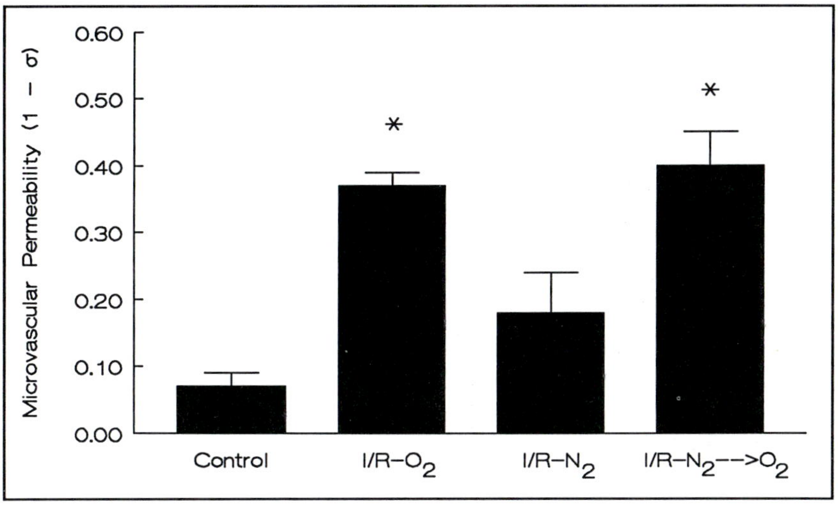

*Fig. 2.1. Ischemia followed by reperfusion with normoxic blood (I/R-O_2) increases microvascular permeability in skeletal muscle. The increase in microvascular permeability is attenuated when ischemic skeletal muscle is reperfused with anoxic blood (I/R-N_2). In the latter experiments, switching from anoxic to normoxic blood (I/R-$N_2 \rightarrow O_2$) for reperfusion induces microvascular barrier disruption (increases permeability). σ = solvent drag reflection coefficient. An * indicates statistical significance from control (nonischemic) preparations. Figure modified from Korthuis et al.[8]*

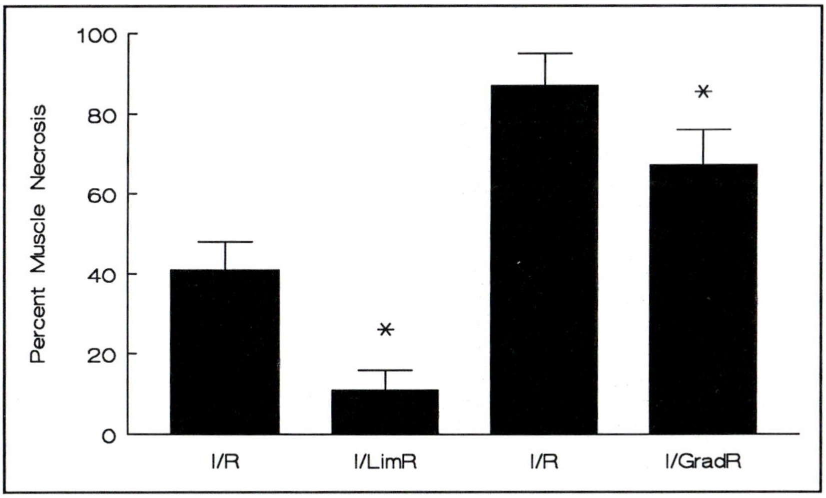

*Fig. 2.2. Necrosis is reduced in postischemic skeletal muscle when the rate of blood flow during reperfusion is limited to preischemic rates (I/LimR) and by reperfusion with a perfusate to which oxygen and erythrocytes are gradually added during reperfusion (I/GradR). An * indicates statistical significance from I/R. Figure modified from Walker et a1[46] and Wright et al.[10]*

and is initiated by the reintroduction of molecular oxygen when blood flow is reestablished.

Protection Afforded by Free Radical Scavengers

The observation that limiting the availability of molecular oxygen at reperfusion prevents the tissue injury and dysfunction induced by reinstitution of flow suggests that ROM may contribute to the pathogenesis of I/R. More compelling support for this concept is derived from studies in which oxidant scavengers were administered prior to reperfusion. The reduction in tissue injury afforded by antioxidant treatment suggests that antioxidants may have applications in the treatment of I/R injury.

Pretreatment with SOD, a superoxide radical scavenging enzyme, has been shown to attenuate postischemic microvascular and parenchymal cell dysfunction in a wide variety of tissues.[21,23,50-57] Catalase (CAT), an enzyme that catalyzes the disproportionation of hydrogen peroxide to water and molecular oxygen, has also proved to be protective in many models of I/R.[21,54] For example, the extent of skeletal muscle necrosis is significantly reduced by gradually reintroducing molecular oxygen during reperfusion and simultaneously administering a combination of SOD, CAT and mannitol.[46] Additionally, treatment with either allopurinol, SOD (Fig. 2.3) or CAT attenuates microvascular barrier disruption in postischemic skeletal muscle and improves capillary perfusion in postischemic skeletal muscle.[21,57] Moreover, calcium uptake by the sarcoplasmic reticulum is preserved by SOD pretreatment in postischemic skeletal muscle.[52] The reduction in resting membrane potential noted during ischemia and reperfusion in rat skeletal muscle is restored to near nonischemic levels by pretreatment with SOD (Fig. 2.3).[51] Finally, the development of compartment syndrome is inhibited by free radical ablation.[58]

One of the major limitations of SOD therapy is its short circulatory half-life in plasma (< 10 min) because it is rapidly excreted by the kidneys (Fig. 2.4).[59-61] Therefore, several methods have been developed to extend the circulating half-life of the enzyme. For example, two human Cu/Zn-SOD units have been linked to the hinge sequence of the immunoglobulin A. This approach substantially reduces the rate of renal excretion and extends the serum half-life to 145 min.[62] Covalent linkage of SOD to polyethylene glycol (PEG) polymer[63,64] has also been used as a means to reduce renal excretion of this antioxidant enzyme. This strategy prolongs the circulatory half-life to 18–40 hours in rats. The use of PEG-SOD is also advantageous because the conjugate is less immunogenic than native SOD, is more resistant to proteolysis, and appears to more readily gain access to the intracellular compartment. A third approach for extending the circulating half-life of SOD involves the incorporation of the enzyme into synthetic liposomes. Like the PEG conjugation approach, liposome encapsulation of SOD substantially increases the circulating half-life (to 4 hours) and facilitates the entry of the enzyme into the cell.[59]

Although the aforementioned modifications effectively increase the circulating half-life of SOD, these approaches may be viewed as a double-edged sword because the increased size of the conjugates reduces their permeation across the microvascular wall. As a consequence, the rate of transport across the microvascular endothelium is reduced, the effect of which is to increase the time it takes to achieve an effective SOD concentration in the muscle interstitium.

A second concern regarding the use of SOD as a therapeutic agent relates to the observation that high doses of SOD fail to confer protection against reperfusion injury. In fact, high doses may actually exacerbate postischemic tissue injury.[65,66] This dose-dependent effect may be related to the fact that in the process of scavenging superoxide, SOD generates hydrogen peroxide, which in turn, can directly injure tissues and may also precipitate the formation of highly reactive hydroxyl radicals. With lower doses of SOD, the hydrogen peroxide formed by the enzyme can be readily eliminated by CAT and other peroxidases. However, the formation of a large amount of superoxide in the presence of excessive quantities of SOD may result in the overproduction of hydrogen peroxide

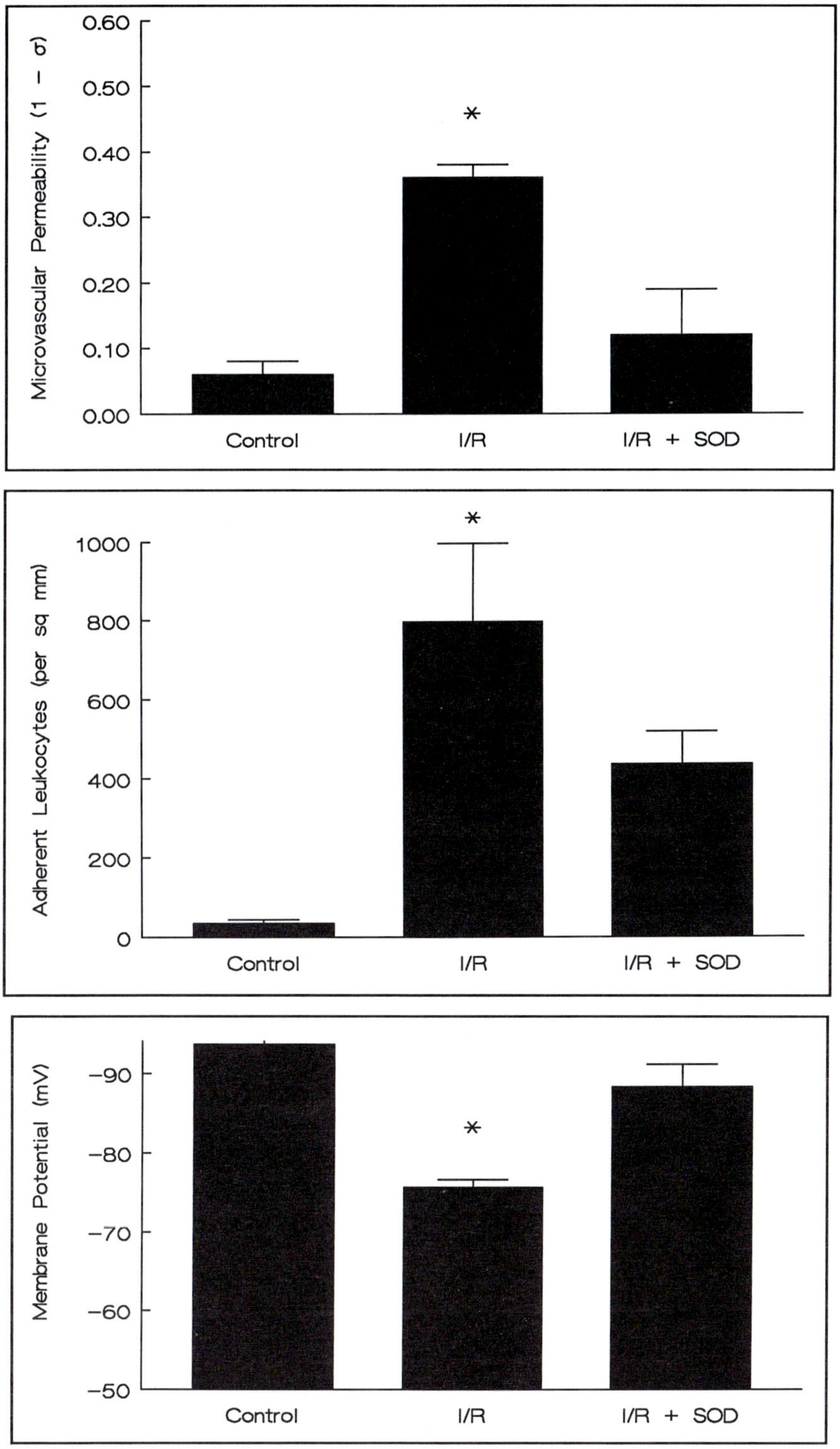

*Fig. 2.3. Superoxide dismutase (SOD) attenuates the increase in microvascular permeability (Upper Panel), leukocyte adherence in postcapillary and collecting venules (Middle Panel), and the reduction in resting membrane potential (Lower Panel) induced by ischemia/reperfusion (I/R). σ = solvent drag reflection coefficient. An * indicates statistical significance from control (nonischemic) preparations. Figure modified from Korthuis et al,[21] Menger et al,[61] and Fantini et al.[51]*

which overwhelms the scavenging capacity of cellular peroxidases.[65,66] This assertion is supported by the observations that administration of SOD-CAT equivalent conjugates, at either low or high doses, is more effective than comparable doses of SOD alone. Moreover, CAT conjugated to SOD entirely removes the toxicity of high dose SOD.[66] These results suggest that the balance between SOD and CAT may be more important than the level of SOD alone.

Nonenzymatic scavengers of the hydroxyl radical, including dimethyl sulfoxide (DMSO), dimethylthiourea (DMTU), mannitol, and others, also protect against reperfusion-induced injury in postischemic skeletal muscle.[21,67] Since hydroxyl radicals are formed from superoxide and hydrogen peroxide, and hydroxyl radical scavengers provide similar levels of protection as SOD or CAT, it has been suggested that the secondarily derived hydroxyl radical may play a major role in the genesis of tissue damage accompanying I/R.[21,68]

Iron Chelators Attenuate Postischemic Skeletal Muscle Dysfunction

In view of the observation that hydroxyl radical scavengers reduce tissue injury in postischemic skeletal muscle and the fact that it has long been known that transition metals such as iron rapidly catalyze the formation of hydroxyl radicals, it is conceivable that iron sequestration may reduce I/R-induced cell dysfunction and necrosis. Indeed, administration of deferoxamine (an iron chelator) or apotransferrin (an iron binding protein) has been shown to attenuate reperfusion injury in many tissues.[37,69-74] In skeletal muscle, pretreatment with deferoxamine attenuates postischemic microvascular barrier disruption,[37] prevents lipid peroxidation[69] and restores the electrical properties of the myocyte cell membrane during reperfusion.[69] More recently, Fantini et al[69] have demonstrated that pentastarch-conjugated deferoxamine reduces sarcolemmal membrane dysfunction. This observation is consistent with the concept that iron chelators exert their effects at extracellular sites in postischemic skeletal muscle. Similar effects have been noted in the heart.[74]

In addition to the ability to bind iron, deferoxamine has been shown to exert other actions, including direct scavenging of free radicals[75-78] and inhibition of MPO activity.[79] While these observations complicate the interpretation of studies using deferoxamine as a probe for the role of iron in I/R, the results of studies which demonstrate that iron-loaded deferoxamine is ineffective in attenuating reperfusion injury[37,71,74] suggest that this is not a major problem. In addition, the fact that apotransferrin, a chemically dissimilar iron-binding agent, exerts a similar protective effect to that noted with deferoxamine treatment in postischemic skeletal muscle[37] supports the likelihood that the protective

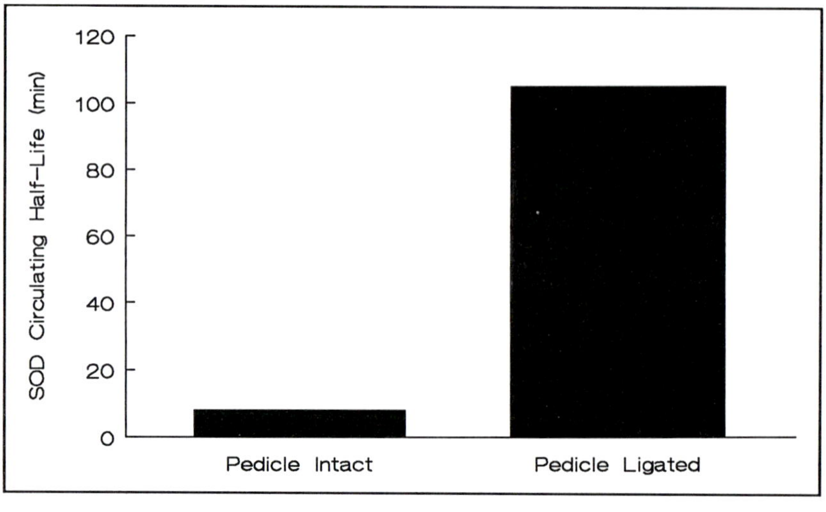

Fig. 2.4. Superoxide dismutase (SOD) is rapidly eliminated (half-life = 8 min) from the plasma in animals in which the renal vascular pedicles supplying the right and left kidneys are intact. Ligation of the renal vascular pedicles significantly extends the circulating half-life (120 min) of SOD. Data from Korthuis et al[61] and Turrens et al.[59]

effect of deferoxamine in postischemic tissues can be primarily attributed to its capacity to bind iron.

Protection Afforded by Attenuation of Oxidant Production

XO has been postulated to play a major role in the genesis of injury produced by reperfusion in ischemic skeletal muscle, where it serves as an important source of ROM. The notion that XO contributes to I/R injury is supported by several lines of evidence. First, XO is a well-documented biological source of ROM, producing both superoxide and hydrogen peroxide.[43] Second, XO is found in skeletal muscle.[15,22,37,39,40] Third, XO activity increases two- to threefold after 2 hours of ischemia in rat skeletal muscle.[22,37] Fourth, hypoxanthine, a major substrate for XO, increases several-fold as a result of ATP catabolism during skeletal muscle ischemia.[30-32] Finally, XO inhibition with allopurinol or oxypurinol[21,22,80] or XO depletion by dietary administration of tungsten[22,25,80,81] significantly attenuates I/R-induced microvascular injury. The deleterious effects of XO-induced oxidant formation have also been reduced by use of an immunoneutralizing polyclonal antibody.[82]

XO-derived ROM may contribute to postischemic tissue injury induced by leukocytes. As will be discussed below, XO-derived oxidants may be involved in the elaboration of chemotactic stimuli in postischemic tissues. As such, XO inhibition reduces the extent of postischemic neutrophil infiltration and attenuates granulocyte-dependent tissue dysfunction. In addition, recent work indicates that XO depletion or inhibition reduces tissue injury induced by neutrophilic elastase[83] and attenuates the lysis of endothelial cell caused by activated neutrophils.[84] These latter studies suggest that XO and activated neutrophils may cooperate in a synergistic fashion to induce tissue injury in postischemic skeletal muscle.

Although the protection afforded by XO inhibitors supports the notion that this enzyme is an important source of ROM in postischemic skeletal muscle, direct radical scavenging by these compounds has been proposed as an alternate explanation.[85] However, careful attention to dose obviates this concern since it has been shown that maintaining extracellular fluid (ECF) concentrations in the range of 10–20 μM produces an 80% reduction in XO activity without enhancing the radical-scavenging potential of ECF.[86] Accordingly, it seems unlikely that the protective actions of these agents can be due to their abilities to scavenge radicals in studies using doses that achieve concentrations in this range.

It should be pointed out that since XO is not present in some species and the enzyme exhibits considerable variability in its activity in different organs in a given species, the XO hypothesis should not be extrapolated to all muscles in any species without proper validation. For example, XO inhibition is ineffective in attenuating injury caused by I/R in the canine gracilis, a muscle which does not contain detectable levels of XO.[33,87,88]

Neutrophils Contribute to the Pathogenesis of Muscle I/R

Leukocyte infiltration into postischemic skeletal muscle has been monitored by the appearance of radiolabeled leukocytes or by measuring the level of granulocyte specific enzymes, such as MPO, in postischemic tissues. Intravital microscopic approaches have also been used to examine this phenomenon. These approaches indicate that reperfusion of ischemic skeletal muscle is associated with leukocyte accumulation (Figs. 2.5 and 2.6).[57,89-92] Given the capacity of phagocytic leukocytes to induce cellular damage to host tissue, the recruitment of neutrophils to postischemic tissues is often cited as circumstantial evidence for the role of granulocytes in the pathogenesis of I/R injury.

More compelling support for the notion that leukocyte infiltration contributes to the pathogenesis of reperfusion injury is provided by studies in which animals are rendered leukopenic by administration of antineutrophil serum, antitumor agents, exposure to X-irradiation or by use of leukopak filters[1,3,93-99] prior to reperfusion. Postischemic microvascular and parenchymal cell dysfunction were attenuated in neutrophil-depleted

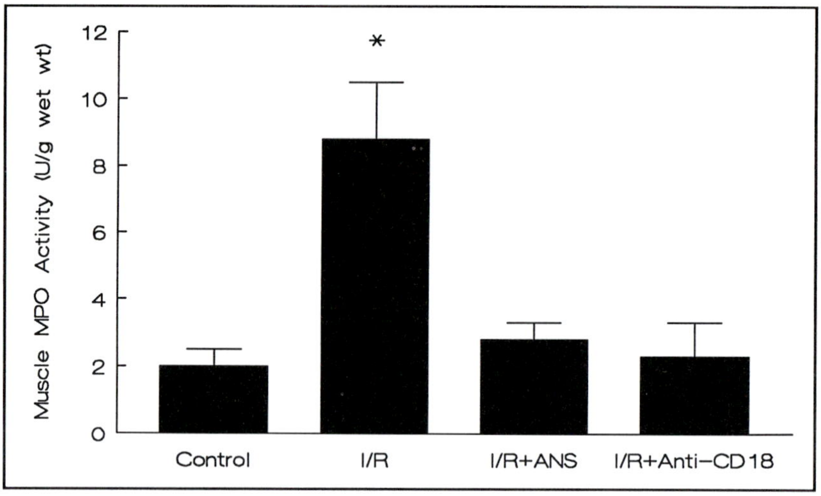

Fig. 2.5. Ischemia/reperfusion (I/R) is associated with marked neutrophil infiltration as indicated by the large increase in skeletal muscle myeloperoxidase (MPO) activity. The postischemic increase in tissue neutrophil content was abolished by prior depletion of circulating neutrophils with antineutrophil serum (ANS) or by administration of an antibody directed against functional epitope on the common β-subunit (CD18) of the neutrophil membrane adhesion glycoprotein CD11/CD18 at the onset of reperfusion. Figure derived from data presented in Carden et al[115] and Carden et al.[87]

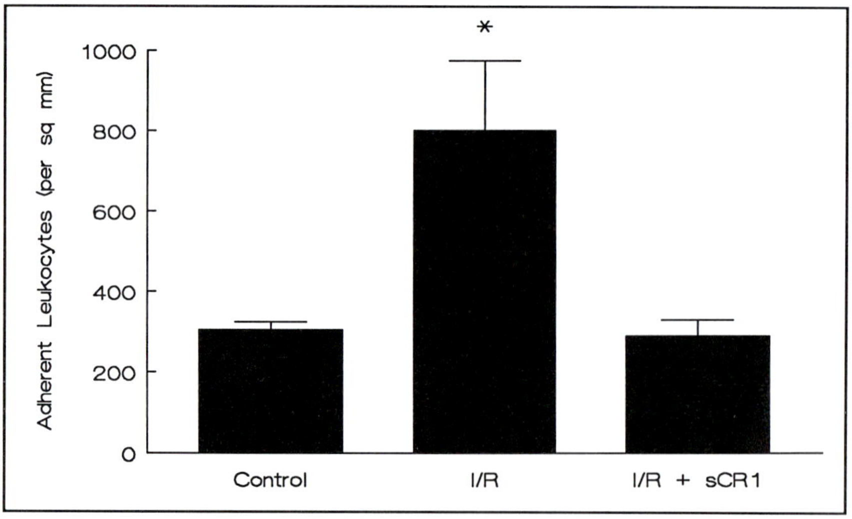

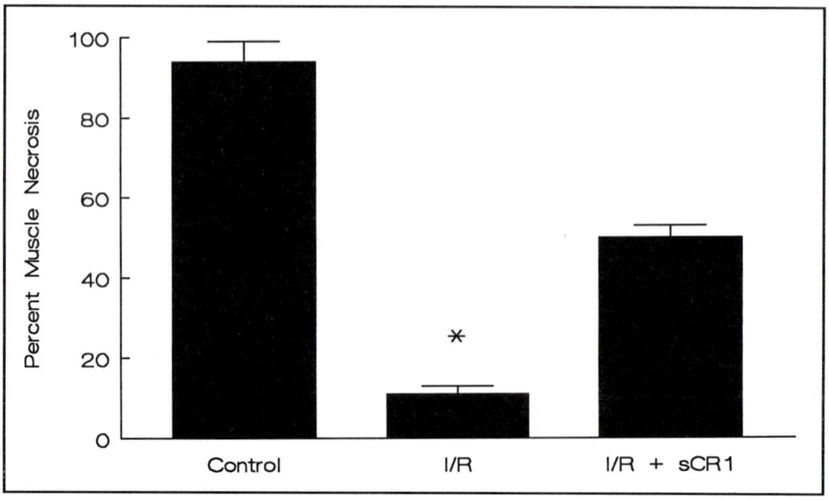

Fig. 2.6. Inhibition of the complement cascade with soluble human recombinant CR1 (sCR1) abolishes ischemia/reperfusion (I/R) induced leukocyte adhesion to postcapillary venules (upper panel) and attenuates necrosis (lower panel) in cremaster muscles of the mouse. Figures derived from data presented in Pemberton et al.[91]

animals (Figs. 2.7 and 2.8). These data suggest that neutrophil accumulation into postischemic tissue is a cause rather than an effect of reperfusion injury.

Since early studies utilizing neutrophil depletion established granulocytes as major contributors to skeletal muscle dysfunction produced by I/R, the potential role for leukocyte adherence to postcapillary venular endothelium in the pathogenesis of reperfusion injury has been evaluated. The rationale for presuming that such an approach would reduce postischemic tissue injury is based on the following observations. Studies involving in vitro models of acute inflammation indicate that close approximation of neutrophils to endothelial cells is required for the production of leukocyte-dependent cellular injury.[44,100-103] Inhibition of leukocyte adherence to venular endothelium also prevents emigration of these inflammatory phagocytes into the tissue space.[100,102,104-106] As a consequence, neutrophil-dependent parenchymal cell injury is reduced. Furthermore, establishment of neutrophil/endothelial cell adhesive interactions not only allows these phagocytic cells to marginate and extravasate at sites of inflammation but also appears to facilitate the destructive potential of the granulocyte. For example, neutrophil/endothelium adherence reactions may establish a "protected microenvironment" that shields granulocyte-derived mediators from inactivation by plasma inhibitors.[44] Establishing an intimate area of contact

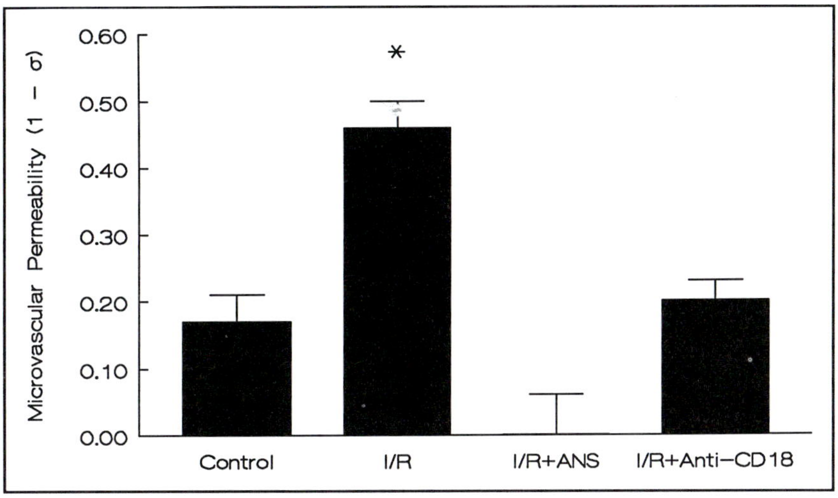

*Fig. 2.7. Ischemia/reperfusion (I/R) induces an increase in skeletal muscle microvascular permeability that is attenuated by prior depletion of circulating neutrophils with antineutrophil serum (ANS) or by administration of an antibody directed against functional epitope on the common β-subunit (CD18) of the neutrophil membrane adhesion glycoprotein CD11/CD18 at the onset of reperfusion. An * indicates statistical significance from control (nonischemic) preparations. Figure derived from data presented in Carden et al.[115]*

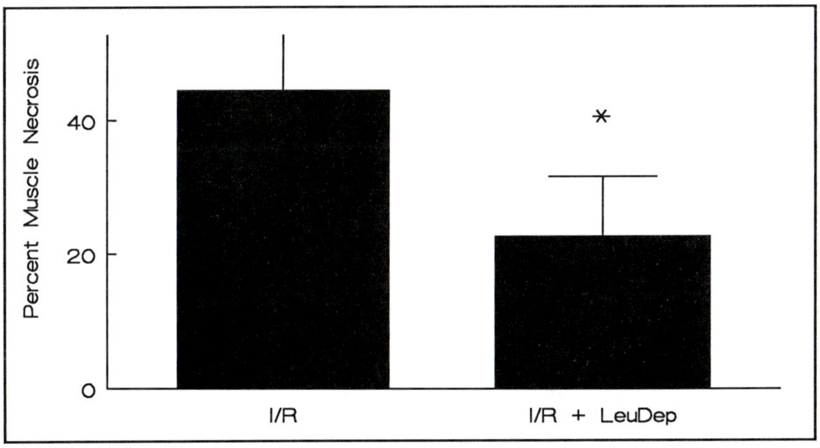

Fig. 2.8. Leukocyte depletion (LeuDep) reduces the extent of skeletal muscle necrosis induced by ischemia/reperfusion (I/R). An asterisk signifies statistical significance at $p<0.05$. Figure modified from Rubin et al.[98]

between granulocytes and target cells also allows the cytotoxic arsenal of neutrophils to be focused at a specific cellular site within this local environment. In addition, adherent neutrophils release far greater quantities of oxidants than suspended cells.[107-110] Moreover, neutrophil adhesion appears to initiate the direct transfer of oxidants from the neutrophil to the intracellular compartment of the cell to which they adhere.[111] Finally, inhibition of leukocyte adherence to endothelial cells or prevention of homotypic aggregation may prevent microvascular occlusion (no-reflow) in postischemic tissues. Thus, it appears that the establishment of adhesive interactions between leukocytes and postcapillary venular endothelium may facilitate the destructive potential of the granulocyte by multiple mechanisms.

The accumulation of neutrophils in postischemic tissues requires that granulocytes first adhere to postcapillary venular endothelium. An important factor in the modulation of the adhesive interaction between granulocytes and venular endothelium is the proadhesive force generated by adhesion molecules expressed on the surface of neutrophils and postcapillary venular endothelium. Whether neutrophils adhere to venular endothelium largely depends on the balance between adhesion forces generated between the leukocyte and endothelium and the hydrodynamic dispersal forces that tend to sweep leukocytes away from the venular wall.[112-114] With regard to adhesion molecules, most of the available evidence implicates the leukocyte integrin, CD11/CD18. This adhesive glycoprotein complex is composed of a family of three structurally and functionally related glycoprotein heterodimers, each consisting of an immunologically distinct α-subunit (CD11) that is noncovalently associated with a common β-subunit (CD18). Antibodies recognizing functional epitopes on the common β-chain have been shown to attenuate I/R-induced neutrophil sequestration,[115-118] neutrophil adherence and emigration in postcapillary venules,[119-122] myocyte contractile dysfunction,[123] microvascular barrier disruption (Fig. 2.8) and capillary no-reflow (Fig. 2.9) in skeletal muscle.[124-126] Although the aforementioned studies clearly establish an important role for the common β-subunit of CD11/CD18 in the pathogenesis of I/R, the contribution of the different α-subunits (CD11a, CD11b, and CD11c) has not been evaluated in postischemic skeletal muscle.

The site specificity of neutrophil adhesion to postcapillary venules in regions of the tissue that were subjected to I/R implies that it involves adhesive structures expressed on the surface of the endothelium in the affected area. Since intercellular adhesion molecule-1 (ICAM-1) is a counter-receptor for CD11/CD18 and its expression can be increased by exposure of the endothelium to many of the bioactive compounds that appear

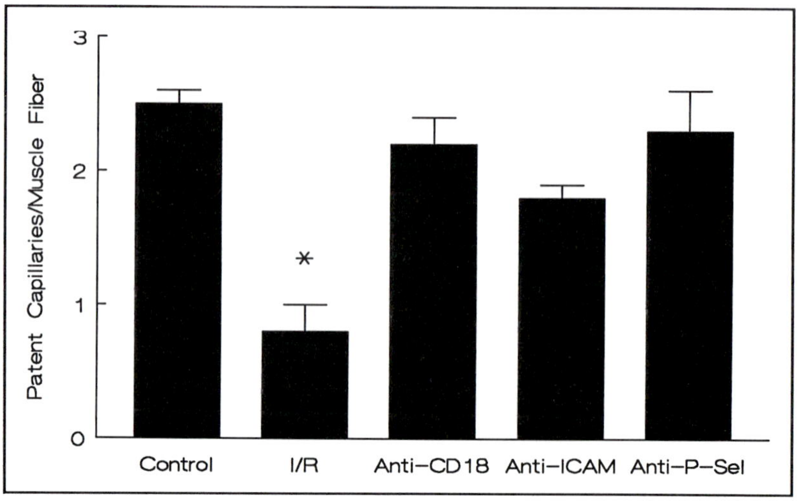

*Fig. 2.9. Immunoneutralization of CD11/C18 (Anti-CD18), intercellular adhesion molecule-1 (Anti-ICAM), or P-selectin (Anti-P-Sel) restores capillary perfusion in skeletal muscles subjected to ischemia/reperfusion (I/R). An * indicates statistical significance from control (nonischemic) preparations. Figure derived from data presented in Jerome et al[175] and Jerome et al.[185]*

to be involved in the pathogenesis of I/R, it has been suggested that ICAM-1 may be an important molecular determinant of neutrophil-endothelial cell adhesion in postischemic tissues.[127-129] The establishment of CD18/ICAM-1-dependent adherence reactions appears to initiate the direct transfer of oxidants from the adherent neutrophil to the intracellular compartment of the myocyte.[130] Thus, ICAM-1-dependent leukocyte adherence to parenchymal cells may enhance targeted cellular destruction by emigrated granulocytes in postischemic tissues. CD18/ICAM-1 adherence reactions contribute to microvascular barrier disruption and the development of postischemic capillary no-reflow in skeletal muscle (Fig. 2.9).[125,126] As a consequence of these microcirculatory events, cellular nutrition is impaired during reperfusion despite the restoration of blood flow.

The role of endothelial E-selectin in the pathogenesis of postischemic skeletal muscle dysfunction has not been evaluated. However, a role for E-selectin in the acute inflammation associated with I/R seems unlikely since the molecule is not constitutively expressed and it requires several hours for full expression after exposure to proinflammatory mediators.[131,132] Nevertheless, it is possible that E-selectin plays an important role in more delayed adherence reactions induced by I/R. Furthermore, since skeletal muscle is very resistant to the short periods of ischemia that are so devastating to other organs such as the heart and brain, the duration of ischemia required to produce injury in the extremities (>4 hr) may allow for changes that ultimately promote E-selectin expression during reperfusion. Indeed, a recent preliminary report indicates that E-selectin antibodies prevent leukocyte rolling and microvascular barrier disruption in postischemic skeletal muscle.[122]

If the mass of skeletal muscle subjected to I/R is sufficiently large, neutrophil-dependent injury can occur in organ sites distant from the tissue subjected to the ischemic insult. For example, the remote consequences of I/R affecting large portions of the lower limbs or the small intestine include neutrophil sequestration in the liver and lungs and the development of hepatic injury and noncardiogenic pulmonary edema.[133-136] This remote injury appears to be neutrophil-dependent since granulocyte depletion effectively eliminates the formation of pulmonary edema induced by lower torso ischemia.[135] Moreover, immunoneutralization of CD11/CD18 eliminates the disruption of the pulmonary microvascular barrier and release of liver enzymes induced by hindlimb I/R, indicating that neutrophil adhesion to microvascular endothelium plays a very important role in distant site injury.[136] Further support for this notion is provided by the observation that P-selectin antibodies prevent the pulmonary complications induced by intestinal I/R.[137]

Although the mechanisms responsible for the selective accumulation of neutrophils in the lungs after I/R of large muscle masses are not clear, it seems likely that proinflammatory mediators produced in the postischemic limb muscles may upregulate the expression of adhesion structures on neutrophils and induce their sequestration in the first microvascular bed they encounter after activation.[4] In addition, it is possible that significant plasma concentrations of proinflammatory mediators, e.g., thromboxane and cytokines, can be achieved if the mass of tissue subjected to I/R is large, as in lower torso ischemia, a notion supported by the observation that postischemic plasma induces neutrophil infiltration when injected into nonischemic skin or dermabrasion chambers.[138,139] As a consequence, these agents may directly activate the endothelium in organs distant to the ischemic site to become adhesive for neutrophils.

The accumulation of neutrophils in postischemic tissues is well-documented and it is assumed that these extravasated leukocytes are responsible for the production of parenchymal cell injury. Recent studies indicate that alterations in the endothelial cell cytoskeleton play an important role in modulating neutrophil accumulation in inflamed tissues. For example, it has been elegantly demonstrated that microfilament-rich endothelial cell lamellopodia extend luminally and envelope leukocytes as they emigrate.[140] In addition, stabilization of F-actin in endothelial cells with phalloidin prevents the alterations in cytoskeletal architecture and leukocyte emigration that occurs in response to a variety of proinflammatory agents both in vitro and in vivo.[105,141,142] Recent studies indicate that postischemic microvascular dysfunction and neutrophil sequestration can be attenuated by stabilization of microfilamentous cytoskeletal elements in the endothelium (Fig. 2.10).[143] Intravital microscopic studies indicate that phalloidin does not alter the increased leukocyte adherence to venular endothelium induced by platelet activating factor (PAF), leukotriene B_4 (LTB_4) and N-formylmethionyl-leucyl-phenylalanine (FMLP)[144] or I/R.[192] However, this agent completely prevented leukocyte emigration and increased vascular albumin clearance induced by these proinflammatory stimuli[144] or I/R.[192] Taken together, these studies suggest that prevention of leukocyte emigration by use of compounds which stabilize the architecture of the endothelial cell cytoskeleton may represent a useful new therapeutic approach to ameliorate I/R injury.

REACTIVE OXYGEN METABOLITES PROMOTE POSTISCHEMIC LEUKOCYTE INFILTRATION

Although it is well established that reperfusion promotes the recruitment of leukocytes into postischemic tissues, the nature of the chemotactic signal remains uncertain. However, several chemical mediators for this process, including ROM, complement, PAF and LTB_4 have been proposed. The notion that ROM participate in the leukocyte recruitment is based on several lines of evidence. For example, endothelial cell monolayers or isolated vessels exposed to hydrogen peroxide demonstrate increased adhesivity for neutrophils.[145,146] In addition, superoxide can inactivate NO, an endogenously produced antiadhesive molecule.[12] More compelling support is provided by the observations that administration of XO inhibitors, SOD, CAT, DMTU (a hydroxyl radical scavenger), or deferoxamine significantly attenuates neutrophil accumulation caused by I/R.[24,42,57,93,147-149] Although SOD also prevents the adherence of leukocytes to cultured endothelial cell monolayers exposed to anoxia/reoxygenation, adherence of activated neutrophils to a biologically inert surface (plastic) was not affected by this enzyme.[120] This result suggests that the modulation of leukocyte adherence by superoxide occurs by an endothelium-dependent mechanism in postischemic tissues.

While the mechanism by which ROM mediate reperfusion-induced leukocyte recruitment is not apparent, several potential explanations are suggested by evidence in the literature. One possibility is that ROM may

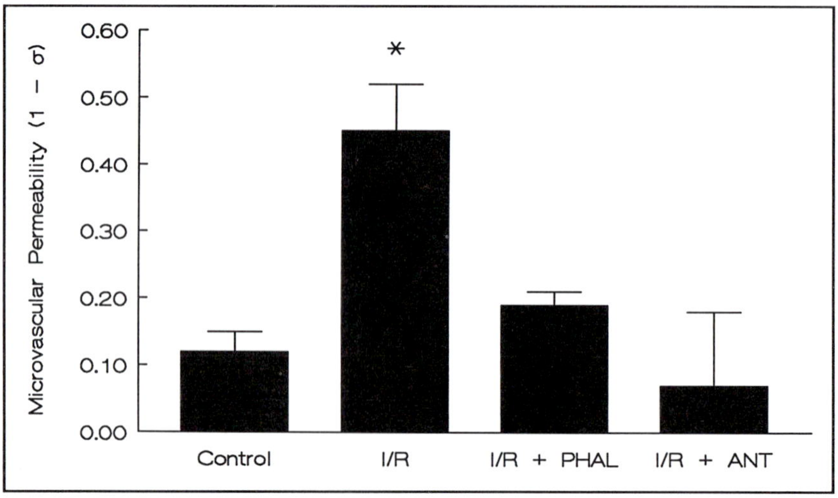

*Fig. 2.10. Stabilization of microfilamentous cytoskeletal elements in endothelium with phalloidin (PHAL) or antamanide (ANT) prevents the increase in skeletal muscle microvascular permeability induced by ischemia/reperfusion (I/R). σ = solvent drag reflection coefficient. An * indicates statistical significance from control (nonischemic) preparations. Figure modified from Korthuis et al.*[143]

stimulate the synthesis and/or release of leukocyte chemoattractants from endothelial cells. For example, exposure of endothelial cells to oxidants can activate phospholipase A_2, resulting in the production of potent leukocyte chemoattractants and activators including PAF and LTB_4.[121] This possibility is supported by the observation that PAF receptor antagonists significantly attenuate hydrogen peroxide-induced leukocyte/endothelial cell adhesion both in vitro and in vivo.[121,145] Moreover, plasma exposed to hydrogen peroxide in the presence of either hemoglobin or iron induces significant neutrophil infiltration.[147] In addition, intradermal injection of plasma obtained from blood draining the gracilis muscle during reperfusion provokes neutrophil infiltration into the injection sites.[139] These studies indicate that a chemotactic signal is formed or released into the bloodstream during reperfusion. Decomplementing postischemic plasma prior to dermal injection markedly reduces neutrophil infiltration into the injection sites.[139] In addition, plasma obtained during reperfusion of SOD-treated muscles also fails to induce significant neutrophil infiltration after intradermal injection.[139] These results suggest that oxidant-induced complement activation may play a role in recruiting neutrophils to postischemic skeletal muscle, a notion supported by the observation that complement activation can be achieved by oxidant exposure in vitro.[150,151] Further support for this concept is provided by the observations that postischemic neutrophil infiltration and muscle injury is prevented by complement blockade with soluble recombinant CR1 (Fig. 2.6) and by prior administration of SOD (Fig. 2.3).[91,139]

A second potential mechanism for oxidant-mediated neutrophil accumulation is that ROM may inactivate endogenous antiadhesive molecules, such as NO. NO is produced by a variety of cells including vascular endothelial cells and neutrophils, and is rapidly inactivated by superoxide. Since several recent studies demonstrated that NO is an important endogenous inhibitor of leukocyte adhesion to vascular endothelium,[152-154] it is conceivable that in postischemic tissue, superoxide generated by endothelial cells and/or leukocytes inactivates NO, resulting in the promotion of leukocyte adhesion.

Another likely explanation for the ability of oxidants to induce leukocyte adhesion relates to the possibility that ROM may directly induce the expression of neutrophil and/or endothelial cell adhesion molecules. The observations that hydrogen peroxide induces endothelial cell P-selectin expression while isolated neutrophils incubated in the presence of hydrogen peroxide or NH_2Cl increase their expression of CD11/CD18 support this possibility.[155,156]

MECHANISMS OF OXIDANT-MEDIATED CELLULAR INJURY

ROM can attack the entire array of biomolecules found in tissues, including structural, contractile, and transport proteins, enzymes, receptors, nucleic acids, membrane lipids and glycosaminoglycans.[12,16] The deleterious molecular effects of reperfusion-induced oxidant production include base hydroxylation, cross-linking, and/or scission of DNA strands, peroxidation of membrane lipids, and cross-linking and degradation of proteins. Reperfusion-induced contractile dysfunction may result as a consequence of oxidant-mediated damage to structural, transport, and contractile proteins and inactivation of receptors and essential enzymes. Peroxidation of membrane lipids may also contribute to myocyte dysfunction as a consequence of alterations in membrane fluidity and porosity, second messenger function, cellular compartmentation, cell volume regulation, the expression and/or mobility of membrane receptors and other proteins, and ultimately, cell lysis.[157,158] As an example, peroxidation of linolenic acid in the acyl chains of cardiolipin, which is essential for the activity of cytochrome oxidase, may result in a loss of function of this critical cellular enzyme.[157]

Another well characterized example of the effects of reperfusion-induced lipid peroxidation relates to the influence of oxidant production on cell membrane potentials. Ischemia has been shown to induce a depolarization of the resting myocyte membrane that persists during reperfusion despite restoration of cellular high-energy phosphate stores.[51,69,97,159,160] Although prior leukocyte depletion or administration of free radical scavengers or the iron chelator deferoxamine did not prevent the ischemia-induced depolarization of the resting myocyte membrane, these interventions were associated with a restoration of resting membrane electrical properties during reperfusion. However, XO inhibition did not provide similar protection.[97] When taken together with the observations that leukocyte depletion or antioxidant therapy prevents membrane lipid peroxidation (as evidenced by an attenuation in the increase in hydroxyconjugated dienes levels in postischemic muscle), the aforementioned results indicate that reperfusion-induced alterations in resting myocyte membrane potential can occur via a neutrophil-dependent, oxidant-mediated mechanism.

The alteration of membrane permeability to ions may also be accompanied by water influx, resulting in cell edema. Moreover, cellular membrane injury caused by oxidants may increase Ca^{2+} influx associated with loss of intracellular K^+, and a decrease in Ca^{2+}-uptake by sarcoplasmic reticulum has been demonstrated in ischemic and reperfused skeletal muscle.[52] These phenomena lead to an increase in intracellular Ca^{2+} concentration, which in turn activate intracellular Ca^+-dependent proteases, phosphatases, and phospholipases, leading to cell injury and still more generation of ROM.

ROM also appear to play a role in the increased microvascular permeability associated with reperfusion of ischemic skeletal muscle. This notion is supported by the observation that leukocyte depletion or the administration of free radical scavengers, XO inhibitors (in some, but not all models), and iron chelators ameliorates the increase in microvascular permeability produced by I/R.[21,22,46,96,161] It is possible that there is a synergistic cytotoxic effect between oxidants and other mediators. For example, oxidants may be involved in the formation and elaboration of proinflammatory mediators by leukocytes or vascular endothelium, as discussed above. Furthermore, the dual exposure of cultured endothelial cells to hydrogen peroxide and proteases promotes much greater cell injury when compared with that noted during exposure to either to hydrogen peroxide or to protease.[162] The latter results indicates that oxidants and proteases interact in a synergistic manner to promote cell destruction.

Neutrophil-derived HOCl also contributes to cellular destruction with I/R. HOCl is a potent oxidizing and chlorinating agent that reacts rapidly with primary amines to yield N-chloramines, which possess the two oxidizing equivalents of hydrogen peroxide and HOCl.[14-16,44] HOCl and N-chloramines cause sulfhydryl oxidation, cytochrome inactivation, chlorination of purine bases on DNA, and degradation of amino acids and proteins, leading to cell damage. For example, it has been shown that HOCl can induce extensive cross-linking of extracellular matrix proteins, e.g., fibronectin, and increase the susceptibility to attack by elastase.[163] Neutrophil-derived HOCl and hydrogen peroxide can also initiate proteolytic reactions catalyzed by neutrophilic proteases. HOCl is able to alter the balance between tissue proteases and antiprotease either by oxidative inactivation of several plasma inhibitors, e.g., α-1-antiprotease, or by oxidative activation of precursor forms of proteases, e.g., collagenase, gelatinase, or both.[16,44,164,165] As α-1-antiprotease protects tissue elastin from hydrolysis by elastase, inactivating α-1-antiprotease enhances tissue injury caused by I/R. Activation of collagenase or gelatinase results in breakdown of connective tissue matrix. Other granular enzymes released from activated neutrophils, such as elastase, collagenase, and cathepsin G, have been shown to play an important role in I/R-induced injury by degrading the capillary basement membrane and endothelial cell glycocalyx. For example, inhibition of neutrophilic elastase prevents microvascular dysfunction caused by I/R.[166]

POSTISCHEMIC CAPILLARY NO-REFLOW

Reperfusion of ischemic skeletal muscle is associated with the development of the no-reflow phenomenon. That is, upon reinstitution of blood flow, a large proportion of the capillaries fail to reperfuse.[92,96,115,167-180] While this microvascular perfusion defect has been shown to occur in postischemic brain, kidney, heart, and skeletal muscle,[92,96,115,167-172,174-180] the mechanisms underlying the development of the no-reflow phenomenon remain unclear. Although, it has been suggested that microvascular thrombus formation may contribute to this capillary perfusion deficit,[181] intravital microscopic studies suggest that this is not the case since microvessel thrombosis is rarely observed.[182] These in vivo observations are corroborated by both light and electron microscopic examination in that platelet or fibrin thrombi are rarely detected in postischemic skeletal muscle.[180,183] Finally, heparin treatment is not effective in restoring capillary perfusion after I/R.[180]

Several lines of evidence indicate that activated neutrophils are significant contributors to the development of postischemic capillary no-reflow.[96,115,168-172,175,177-179,184] For example, a strong correlation has been noted between the percent of capillaries exhibiting no-reflow and the number of leukocytes present in these capillaries in postischemic tissues.[168-172] More compelling support for this notion is provided by the observation that neutrophil depletion virtually abolishes no-reflow in reperfused myocardium, brain and skeletal muscle.[115,168,175] It appears that oxidants are involved in the impaired capillary perfusion and enhanced leukocyte adhesion since treatment with allopurinol and SOD restore capillary perfusion and inhibit leukocyte/endothelial cell adhesion. The fact that antioxidant treatment prevents both neutrophil/endothelial cell adhesion and improves no-reflow supports the concept that these events may be linked.[57]

While the evidence presented thus far indicates that leukocytes are involved in the development of the no-reflow phenomenon in postischemic tissues, it still remains uncertain as to how neutrophils cause this perfusion defect. However, a growing body of evidence indicates that neutrophil/endothelial cell adhesive interactions are required for the development of capillary no-reflow in postischemic skeletal muscle. For example, immunoneutralization of the common β-subunit (CD18) of the neutrophil membrane CD11/CD18 glycoprotein adherence complex reduces the extent of no-reflow in postischemic brain and skeletal muscle (Fig. 2.9).[115,122,175,178] More recent studies indicate that administration of antibodies directed against functional epitopes on ICAM-1 or P-selectin also attenuates the development of this postischemic perfusion defect (Fig. 2.9).[122,185] Since P-selectin mediates platelet-platelet and platelet-leukocyte adhesion, it is possible that immunoneutralization of this adhesion ligand prevents no-reflow by a mechanism that involves these types of adhesion interactions (Fig. 2.11).

Although the aforementioned studies establish that there is a role for CD11/CD18, ICAM-1, and P-selectin in the development of the no-reflow phenomenon in postischemic skeletal muscle, it is not clear how neutrophils that become adherent to postcapillary venules cause capillary no-reflow. One potential mechanism may be that lumenal obstruction occurs as a direct result of neutrophil adherence to venular endothelium. Although complete obstruction of the venular lumen by adherent leukocytes does not appear to occur in postischemic skeletal muscle,[175,186] partial occlusion by adherent leukocytes may alter microhemodynamics such that flow in upstream vessels, i.e., capillaries, ceases.[187] Physical impaction of leukocytes in the capillary lumen has also been proposed as a mechanism to explain capillary malperfusion during ischemia.[172] However, leukocyte/capillary plugging does not appear to play a role in the pathogenesis of no-reflow when normal perfusion pressures are established at reperfusion.[188] It is also possible that oxidants produced by activated leukocytes disrupt the mechanisms involved in the regulation of endothelial cell volume in postischemic skeletal muscle. As a consequence, endothelial cells swell and reduce the caliber of the capillary lumen,[189,190] thereby contributing to no-reflow. Endothelial cell swelling occurs during ischemia but is potentiated during re-perfusion.[190] Treatment with a hypertonic, hyperosmotic saline dextran solution prevents endothelial cell swelling[191] and reduces the extent of postischemic no-reflow.[192] Although postischemic endothelial cell swelling appears to result from altered Na^+-H^+ exchange,[193] the effect of oxidants or neutrophil activation on Na^+-H^+ exchange in endothelium is not known.

A more likely explanation for the participation of leukocyte adhesion in the genesis of no-reflow may be related to the ability of these cells to induce microvascular barrier disruption (Fig. 2.11). That is, as neutrophils adhere to the postcapillary venular endothelium and emigrate into the tissue spaces, microvascular permeability is increased. As a consequence of these events, transcapillary fluid filtration rate is enhanced which leads to an accumulation of excessive interstitial fluid volume. The rate of edema formation may be further enhanced by an increase in capillary pressure that results as a consequence of the physical reduction in the diameter of postcapillary venules induced by leukocytes adhering to the endothelium in these vessels.[194] Because many muscles cannot readily expand due to enclosure by a restrictive fascial sheath, I/R-induced, neutrophil-mediated edema formation can be accompanied by a rise in interstitial fluid pressure sufficient to produce significant extravascular compression. Since the diameter and intravascular pressure in capillaries is less than in other vessel segments, the capillaries (and venules) would be the first to exhibit no-reflow.

According to the scenario outlined in Figure 2.11 leukocyte adherence does not directly result in the development of no-reflow. Rather the neutrophil-dependent increase in microvascular permeability and edema formation produce capillary no-reflow by a mechanism that may involve extravascular compression. Neutrophil-induced microvascular barrier disruption and altered membrane ion fluxes may also contribute to the development of no-reflow by increasing microvessel hematocrit.[195] As a consequence, microvascular blood fluidity is reduced which may impair capillary perfusion in postischemic skeletal muscle. Support for this notion is derived from studies in which prophylactic isovolemic hemodilution was used to restore hematocrit to preischemic levels[196] and capillary perfusion was restored.

Acknowledgment

This work was supported by a grant from the National Heart, Lung, and Blood Institute (HL-48646). Dr. Korthuis is an Established Investigator of the American Heart Association.

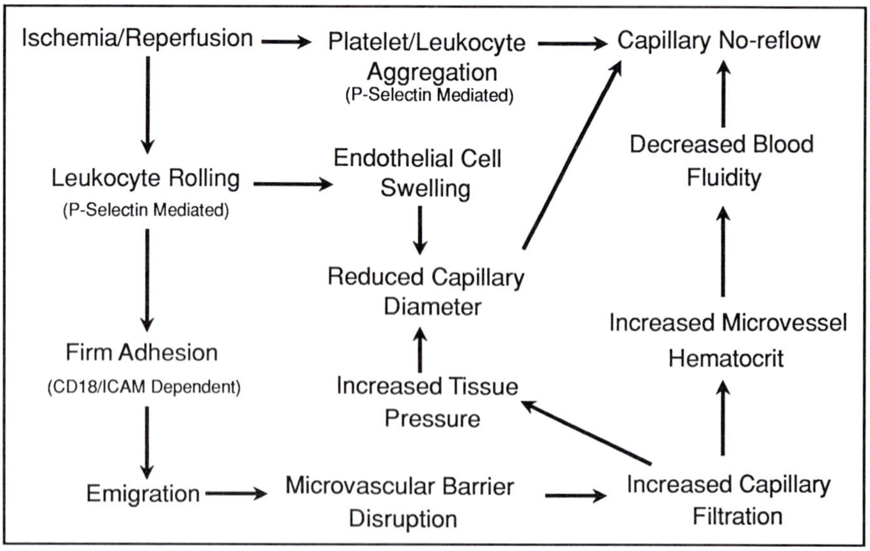

Fig. 2.11. Mechanism proposed to explain the role of leukocyte adhesion to postcapillary venules in producing capillary no-reflow in postischemic skeletal muscle. According to this scenario, ischemia/reperfusion induces P-selectin dependent leukocyte rolling. The rolling leukocytes may then establish stationary adhesive interactions that are mediated by CD18/ICAM-1 dependent mechanisms. Leukocyte emigration disrupts the microvascular barrier and produces an increase in microvascular permeability. As a consequence, transmicrovascular fluid filtration is increased and edema forms in the interstitial space of postischemic skeletal muscle. The excessive accumulation of fluid in the tissue spaces raises interstitial fluid pressure which, in turn, reduces vascular transmural pressure and compresses the capillaries thereby producing capillary no-reflow.

References

1. Carden DL, Korthuis RJ. Mechanisms of postischemic vascular dysfunction in skeletal muscle: implications for therapeutic intervention. Microcirc Endo Lymph. 1989;5:277-98.
2. Lindsay T, Romaschin A, Walker PM. Free radical mediated damage in skeletal muscle. Microcirc Endo Lymph. 1989;5:157-70.
3. Korthuis RJ, Granger DN. Cellular dysfunction induced by ischemia/reperfusion: Role of reactive oxygen metabolites and granulocytes. In: Bloom AD, Spatz L, eds.Biological consequences of oxidative stress: Implications for Cardiovascular disease and carcinogenesis. New York: Oxford University Press, 1992:50-77.
4. Gute DC, Korthuis RJ. Role of leukocyte adherence in reperfusion-induced microvascular dysfunction and tissue injury. In: Granger DN, Schmid-Schoenbein G, eds.Physiology and pathophysiology of leukocyte adhesion. Oxford: Oxford University Press, 1994.
5. Korthuis RJ, Anderson DC, Granger DN. Role of neutrophil-endothelial cell adhesion in inflammatory disorders. J Crit Care. 1994;9(1):47-71.
6. Dahlback LO, Rais O. Morphological changes in striated muscle following ischemia: immediate postischemic phase. Acta Chir Scand. 1966;131:430-8.
7. Koller A, Kaley G. Prostaglandins mediate arteriolar dilation to increased blood flow velocity in skeletal muscle microcirculation. Circ Res. 1990;67(2):529-34.
8. Korthuis RJ, Smith JK, Carden DL. Hypoxic reperfusion attenuates postischemic microvascular injury. Am J Physiol. 1989;256:H315-9.
9. Hearse DJ, Humphrey RM, Chain EB. Abrupt reoxygenation of the anoxic potassium arrested rat heart: A study of myocardial enzyme release. J Mol Cell Cardiol. 1973;5:395-407.
10. Wright JG, Fox D, Kerr JC et al. Rate of reperfusion blood flow modulates reperfusion injury in skeletal muscle. J Surg Res. 1988;44:754-9.
11. Halliwell B, Gutteridge JMC. Free radicals in biology and medicine. Oxford: Clarenden Press, 1989.
12. Grisham MB. Reactive metabolites of oxygen and nitrogen in biology and medicine. Austin, Tx.: R.G.Landes Company, 1993.
13. McCord JM, Fridovich I. Superoxide dismutase. An enzymic function for erythroprien (hemocuprien). J Biol Chem. 1969; 244:6049-55.
14. Grisham MB, Granger DN. Metabolic sources of reactive oxygen metabolites during oxidant stress and ischemia and reperfusion. Clin Chest Med. 1989;10:71-81.
15. Grisham MB, McCord JM. Chemistry and cytotoxicity of reactive oxygen metabolites. In: Taylor AE, Matalon S, Ward PA, eds. Physiology of oxygen radicals. Bethesda: American Physiological Society, 1986:1-18.
16. Weiss SJ. Oxygen, ischemia, and inflammation. Acta Physiol Scand (Suppl). 1986;548:9-38.
17. Beckman JS, Beckman TW, Chen J et al. Apparent hydroxyl radical production by peroxynitrite: implications for endothelial injury from nitric oxide and superoxide. Proc Natl Acad Sci USA. 1990;87:1620-4.
18. Ramos CL, Par S, Britigan BE et al. Spin trapping evidence for myeloperoxidase-dependent hydroxyl radical formation by human neutrophils and monocytes. J Biol Chem. 1992;267:8307-12.
19. Paller MS. Hemoglobin- and myoglobin-induced acute renal failure in rats: role of iron in nephrotoxicity. Am J Physiol. 1988;255:F539-44.
20. Parks DA, Granger DN. Xanthine oxidase: biochemistry, distribution, and physiology. Acta Physiol Scand (Suppl). 1986; 548:87-99.
21. Korthuis RJ, Granger DN, Townsley MI et al. The role of oxygen-derived free radicals in ischemia-induced increases in canine skeletal muscle vascular permeability. Circ Res. 1985;57:599-609.
22. Smith JK, Carden DL, Korthuis RJ. Role of xanthine oxidase in postischemic microvascular injury in skeletal muscle. Am J Physiol. 1989;257:H1782-9.
23. Granger DN, Rutili G, McCord JM. Superoxide radicals in feline intestinal ischemia. Gastroenterology. 1981;81:22-9.
24. Granger DN, McCord JM, Parks DA et al. Xanthine oxidase inhibitors attenuate ischemia-induced vascular permeability changes in the cat intestine. Gastroenterol 1986;90:80-4.

25. McKelvey TG, Hollworth ME, Granger DN et al. Mechanisms of conversion from xanthine dehydrogenase to xanthine oxidase in ischemic rat liver and kidney. Am J Physiol. 1988;254:G753-60.
26. Parks DA, Williams TK, Beckman JS. Conversion of xanthine dehydrogenase to oxidase in ischemic rat intestine: a reevaluation. Am J Physiol. 1988;254:G768-74.
27. Battelli MG. Enzymatic conversion of rat liver xanthine oxidase from dehydrogenase (D form) to oxidase (O form). FEBS Lett. 1980;113:47-51.
28. McCord JM. Free radicals and myocardial ischemia: overview and outlook. Free Radic Biol Med. 1988;4:9-14.
29. Saugstad OD. Hypoxanthine as an indicator of hypoxia: its role in health and disease through free radical production. Pediatr Res. 1988;23:143-50.
30. Gidlof A, Hammersen F, Larsson J et al. Endothelium in human skeletal muscle an ischemic shock tissue. In: Lewis DH, ed.Induced skeletal muscle ischemial in man. Basel: Karger, 1982:63-79.
31. Larsson J, Gidlof A, Lewis DH et al. Effect of induced ischemia on plasma hypoxanthine levels in man. In: Lewis DH, ed. Induced skeletal muscle ischemia in man. Basel: Karger, 1982:49-54.
32. McCord JM. Oxygen derived free radicals in postischemic tissue injury. N Engl J Med. 1985;312:159-63.
33. Lindsay T, Liauw TS, Romaschin AD et al. The effect of ischemia/reperfusion on adenine nucleotide metabolism and xanthine oxidase production in skeletal muscle. J Vasc Surg. 1990;12:8-15.
34. Della Corte E, Stirpe F. The regulation of rat liver xanthine oxidase. Involvement of thiol groups in the conversion of the enzyme activity from the dehydrogenase (type D) into oxidase (type O) and purification of the enzyme. Biochem J. 1972;126:739-45.
35. Waud WR, Rajogopalan KV. Purification and properties of the NAD-dependent (type D) and O_2-dependent forms of rat liver xanthine dehydrogenase. Arch Biochem Biophys. 1988;172:354-64.
36. Parks DA, Granger DN, Bulkley GB et al. Soybean trypsin inhibitor attenuates ischemic injury to the feline small intestine. Gastroenterology. 1985;89:6-12.
37. Smith JK, Carden DL, Grisham MB et al. Role of iron in postischemic microvascular injury. Am J Physiol. 1989;256:H1472-7.
38. Ibrahim B, Stoward PJ. The histochemical localization of xanthine oxidase. Histochem J. 1978;10:615-7.
39. Jarasch ED, Grund C, Bruder G et al. Localization of xanthine oxidase in mammary gland epithelium and capillary endothelium. Cell. 1981;25:67-82.
40. Jarasch ED, Bruder G, Heid HW. Significance of xanthine oxidase in capillary endothelial cells. Acta Physiol Scand (Suppl). 1986;548:39-46.
41. Zweier JL, Kuppusamy P, Lutty GA. Measurement of endothelial cell free radical generation: evidence for a central mechanism of free radical injury in postischemic tissues. Proc Natl Acad Sci USA. 1988; 85:4046-50.
42. Grisham MB, Hernandez LA, Granger DN. Xanthine oxidase and neutrophil infiltration in intestinal ischemia. Am J Physiol. 1986;251:G567-74.
43. Bishop CT, Mirza Z, Freeman BA. Free radical damage to cultured porcine aortic endothelial cells and lung fibroblasts: modulation by culture conditions. In Vitro Cell Dev Biol. 1988;21:229-33.
44. Weiss SJ. Tissue destruction by neutrophils. N Engl J Med. 1989;320:365-76.
45. Winterbourne CC. Myeloperoxidase as an effective inhibitor of hydroxyl radical production. Implications of the oxidative reaction of neutrophils. J Clin Invest. 1986; 78:545-50.
46. Walker PM, Lindsay TF, Labbe R. Salvage of skeletal muscle with free radical scavengers. J Vasc Surg. 1987;5:68-75.
47. Hearse DJ. Reperfusion of the ischemic myocardium. J Mol Cell Cardiol. 1977; 9:605-15.
48. Hearse DJ, Humphrey SM, Nayler WG et al. Ultrastructural damage associated with reoxygenation of anoxic myocardium. J Mol Cell Cardiol. 1975;7:315-24.
49. Perry MA, Wadwa SS. Gradual reintroduction of oxygen reduces reperfusion injury in cat stomach. Am J Physiol. 1988; 254: G366-72.

50. Granger DN, Hollwarth ME, Parks DA. Ischemia-reperfusion injury: role of oxygen-derived free radicals. Acta Physiol Scand. 1986;548:47-63.
51. Fantini GA, Minei JP, Perry MO et al. Reperfusion with superoxide dismutase improves cellular membrane function in postischemic skeletal muscle. Surg Forum. 1987;38:317-9.
52. Lee KR, Cronenwett JL, Shlafer M et al. Effect of superoxide dismutase plus catalase on Ca^{2+} transport in ischemic and reperfused skeletal muscle. J Surg Res. 1987;42:24-32.
53. Grisham MB, Hernandez LA, Granger DN. Xanthine oxidase and neutrophil infiltration in intestinal ischemia. Gastroenterology. 1981;81:22-9.
54. Granger DN, Parks DA. Role of oxygen radicals in the pathogenesis of intestinal ischemia. Physiologist. 1983;26:159-64.
55. Bolli R. Oxygen-derived free radicals and postischemic myocardial dysfunction ("stunned myocardium"). J Am Coll Cardiol. 1988;12:239-49.
56. Menger MD, Steriner D, Messmer K. Microvascular ischemia-reperfusion injury in striated muscle: significance of "no reflow". Am J Physiol. 1992;263:H1892-900.
57. Menger MD, Pelikan S, Steiner D et al. Microvascular ischemia-reperfusion injury in striated muscle: significance of "reflow paradox". Am J Physiol. 1992;263:H1901-6.
58. Perler BA, Tohmen AG, Bulkley GB. Inhibition of the compartment syndrome by ablation of the radical-mediated reperfusion injury. Surgery. 1990;108:40-7.
59. Turrens JF, Crapo JD, Freeman BA. Protection against oxygen toxicity by intravenous injection of liposome-entrapped catalase and superoxide dismutase. J Clin Invest. 1984;73:87-95.
60. Bayati A, Kallskog O, Odlind B et al. Plasma elimination kinetics and renal handling of copper/zinc superoxide dismutase in the rat. Acta Physiol Scand. 1988; 134:65-74.
61. Korthuis RJ, Kubes P, Tso P et al. Transport kinetics for superoxide dismutase and catalase between plasma and interstitial fluid in the rat small intestine. Free Radic Biol Med. 1991;11:293-8.
62. Hallewell RA, Laria RA, Tabrizi A et al. Genetically engineered polymers of human Cu-Zn superoxide dismutase. J Biol Chem. 1989;264:5260-8.
63. Mossman BT, Marsh JP, Hardwick D et al. Approaches to prevention of asbestos-induced lung disease using polyethylene glycol (PEG)-conjugated catalase. Free Radic Biol Med. 1986;2:335-8.
64. Beckman JS, Minor RL, White CW et al. Superoxide dismutase and catalase conjugated to polyethylene glycol increases endothelial enzyme activity and oxidant resistance. J Biol Chem. 1988;263:6884-92.
65. Omar BA, McCord JM. The cardioprotective effect of Mn-superoxide dismutase is lost at high doses in the postischemic isolated rabbit heart. Free Radic Biol Med. 1990; 9:473-8.
66. Mao GD, Thomas PD, Lopaschuk GD et al. Superoxide dismutase (SOD)-catalase conjugates. Role of hydrogen peroxide and the fenton reaction in SOD toxicity. J Biol Chem. 1993;268:416-20.
67. Smith JK, Carden DL, Korthuis RJ. Role of hydroxyl radicals in ischemia/reperfusion-induced injury to skeletal muscle microvasculature. FASEB J. 1989;3:A1234
68. Bolli R, Zhu W, Hartley CJ et al. Attenuation of dysfunction in the postischemic "stunned" myocardium by dimethylthiourea. Circulation. 1987;76:458-68.
69. Fantini GA, Yoshioka T. Deferoxamine prevents lipid peroxidation and attenuates reoxygenation injury in postischemic skeletal muscle. Am J Physiol. 1993; 264: H1953-9.
70. Bolli R, Patel BS, Zhu ZX et al. The iron chelator desferrioxamine attenuates postischemic ventricular dysfunction. Am J Physiol. 1987;253:H1372-80.
71. Hernandez LA, Grisham MB, Granger DN. A role for iron in oxidant-mediated ischemic injury to intestinal microvasculature. Am J Physiol. 1987;253:G49-53.
72. Reddy BR, Kloner RA, Przyklenk K. Early treatment with deferoxamine limits myocardial ischemic/reperfusion injury. Free Radic Biol Med. 1989;7:45-52.
73. Bolli R, Patel BS, Jeroudi MO et al. Iron-mediated radical reactions upon reperfusion contribute to myocardial "stunning". Am J Physiol. 1990;259:H1901-11.

74. Maruyama M, Pieper GM, Kalyanaraman B et al. Effects of hydroxyethyl starch conjugated deferoxamine on myocardial functional recovery following coronary occlusion and reperfusion in dogs. J Cardiovasc Pharmacol. 1991;17:166-75.
75. Sinaceur J, Ribiere C, Nordmann J et al. Desferrioxamine: a scavenger of superoxide radicals. Biochem Pharmacol. 1984; 33: 1693-4.
76. Halliwell B. Use of desferrioxamine as a 'probe' for iron-dependent formation of hydroxyl radicals. Evidence for a direct reaction between desferal and the superoxide radical. Biochem Pharmacol. 1985; 34:229-33.
77. Darley-Usmar VM, Hersey A, Garland LG. A method for comparative assessment of antioxidants as peroxyl radical scavengers. Biochem Pharmacol. 1989;38:1465-9.
78. Hartley A, Davies M, Rice-Evans C. Desferrioxamine as a chainbreaking antioxidant in sickle cell membranes. FEBS Lett. 1990;264:145-8.
79. Klebanoff SJ, Waltersdorph AM. Inhibition of peroxidase-catalyzed reactions by deferoxamine. Arch Biochem Biophys. 1988; 264:600-6.
80. McCutchan HJ, Schwappach JR, Enquist EG et al. Xanthine oxidase-derived H_2O_2 contributes to reperfusion injury of ischemic skeletal muscle. Am J Physiol. 1990; 258:H1415-9.
81. Brown JM, Terada LS, Grosso MA et al. Xanthine oxidase produces hydrogen peroxide which contributes to reperfusion injury of ischemic, isolated, perfused rat hearts. J Clin Invest. 1988;81:1297-301.
82. Terada LS, Dormish JJ, Stanley PF et al. Circulating xanthine oxidase mediates lung neutrophil sequestration after intestinal ischemia-reperfusion. Am J Physiol. 1992; 263:L394-401.
83. Rodell TC, Cheronis JC, Ohnemus CL et al. Xanthine oxidase mediates elastase-induced injury to isolated lungs and endothelium. J Appl Physiol. 1987;63:2159-63.
84. Phan SH, Gannon DE, Varani J et al. Xanthine oxidase activity in rat pulmonary artery endothelial cells and its alteration by activated neutrophils. Am J Pathol. 1989;134:1201-11.
85. Moorhouse PC, Grootveld M, Halliwell B et al. Allopurinol and oxypurinol are hydroxyl radical scavengers. FEBS Lett. 1987;213:23-8.
86. Zimmerman BJ, Parks DA, Grisham MB et al. Allopurinol does not enhance antioxidant properties of extracellular fluid. Am J Physiol. 1988;255:H202-6.
87. Carden DL, Smith JK, Korthuis RJ. Oxidant-mediated, CD18-dependent microvascular dysfunction induced by complement-activated granulocytes. Am J Physiol. 1991;260:H1144-52.
88. Korthuis RJ, Smith JK, Carden DL. Role of xanthine oxidase in postischemic granulocyte infiltration and microvascular injury in canine gracilis muscle. Physiologist (Suppl). 1989;32:158
89. Zimmerman BJ, Guillory DJ, Grisham MB et al. Role of Leukotriene B_4 in granulocyte infiltration into the postischemic feline intestine. Gastroenterology. 1990;99:1358-63.
90. Smith JK, Carden DL, Korthuis RJ. Activated neutrophils increase microvascular permeability in skeletal muscle: role of xanthine oxidase. J Appl Physiol. 1991; 70:2003-9.
91. Pemberton M, Anderson G, Vetvicka V et al. Microvascular effects of complement blockade with soluble recombinant CR1 on ischemia/reperfusion injury in skeletal muscle. J Immunol. 1993;150:5104-13.
92. Lehr HA, Guhlmann A, Nolte D et al. Leukotrienes as mediators in ischemia-reperfusion injury in a microcirculation model in the hamster. J Clin Invest. 1991; 87:2036-41.
93. Granger DN. Role of xanthine oxidase and granulocytes in ischemia-reperfusion injury. Am J Physiol. 1988;255:H1269-75.
94. Granger DN, Villareal D. Mechanisms of reperfusion-induced microvascular dysfunction. J Vasc Surg. 1993;18:104-6.
95. Korthuis RJ, Jerome SN. Role of neutrophils in the pathogenesis of microvascular dysfunction in postischemic skeletal muscle. J Vasc Sug. 1993;18:114-17.
96. Korthuis RJ, Grisham MB, Granger DN. Leukocyte depletion attenuates vascular injury in postischemic skeletal muscle. Am J Physiol 1988;254:H823-7.

97. Yokota J, Minei JP, Fantini GA et al. Role of leukocytes in reperfusion injury of skeletal muscle after partial ischemia. Am J Physiol. 1989;257:H1068-75.
98. Rubin BB, Chang G, Liauw S et al. Phospholipid peroxidation deacylation and remodeling in postischemic skeletal muscle. Am J Physiol. 1992;263:H1695-702.
99. Walden DL, McCutchan HJ, Enquist EG et al. Neutrophils accumulate and contribute to skeletal muscle dysfunction after ischemia-reperfusion. Am J Physiol. 1990; 259:H1809-12.
100. Gallin JI, Goldstein IM, Snyderman R. Inflammation: basic principles and clinical correlates. 2nd ed. NY: Raven Press, 1992:
101. Harlan JM. Leukocyte-endothelial cell interactions. Blood. 1985;65:513-25.
102. Springer T, Anderson DC, Rosenthal A et al. Leukocyte adhesion molecules. NY: Springer-Verlag, 1989:
103. Tonnenson MG. Neutrophil-endothelial cell interactions: mechanisms of neutrophil adherence to vascular endothelium. J Invest Dermatol. 1989;93:53S-8S.
104. Kurtel H, Tso P, Granger DN. Granulocyte accumulation in postischemic intestine: role of leukocyte adhesion glycoprotein CD11/CD18. Am J Physiol 1992; 262:G878-82.
105. Paterson IS, Klausner JM, Goldman G et al. The endothelial cell cytoskeleton modulates extravascular polymorphonuclear leukocyte accumulations in vivo. Microvasc Res. 1989;38:49-56.
106. Tanaka M, Brooks SE, Richard VJ et al. Effect of anti-CD18 antibody on myocardial neutrophil accumulation and infarct size after ischemia and reperfusion in dogs. Circulation. 1993;87:526-35.
107. Entman ML, Youker K, Shappell SB et al. Neutrophil adherence to isolated adult canine myocytes: evidence for a CD18-dependent mechanism. J Clin Invest. 1990; 85:1497-506.
108. Nathan CF. Neutrophil activation on biological surfaces: massive secretion of hydrogen peroxide in response to products of macrophages and lymphocytes. J Clin Invest. 1987;80:1550-60.
109. Nathan CF. Respiratory burst in adherent human neutrophils: triggering by colony stimulating factors CSF-GM and CSF-G. Blood. 1989;73:301-6.
110. Shappell SB, Toman C, Anderson DC et al. Mac-1 (CD11b/CD18) mediates adherence-dependent hydrogen peroxide production by human and canine neutrophils. J Immunol. 1990;144:2702-11.
111. Entman ML, Youker K, Shoji T et al. Neutrophil-induced oxidative injury of cardiac myocytes. A compartmented system requiring CD11b/CD18-ICAM-1 adherence. J Clin Invest. 1992;90:1335-45.
112. Atherton A, Born GVR. Relationship between the velocity of rolling granulocytes and that of blood flow in venules. J Physiol. 1973;233:157-65.
113. Schmid-Schoenbein GW, Shih YY, Chien S. Vascular endothelium-leukocyte interaction, sticking shear force in capillaries. Circ Res. 1975;36:173-84.
114. Mayrovitz HN, Wiedeman MP, Tuma RF. Experimental and clinical-thrombosis: factors influencing leukocyte adherence in microvessels. Thromb Haemostasis. 1977; 38:823-30.
115. Carden DL, Smith JK, Korthuis RJ. Neutrophil-mediated microvascular dysfunction in postischemic canine skeletal muscle: role of granulocyte adherence. Circ Res. 1990; 66:1436-44.
116. Dreyer WJ, Michael LH, West MS et al. Neutrophil accumulation in ischemic myocardium. Insights into time course, distribution, and mechanism of localization during early reperfusion. Circulation 1991; 84:400-11.
117. Horgan MJ, Wright SD, Malik AB. Antibody against leukocyte integrin (CD18) prevents reperfusion-induced lung vascular injury. Am J Physiol. 1990;259:L315-9.
118. Schoenberg MH, Poch B, Younes M et al. Involvement of neutrophils in postischaemic damage to the small intestine. Gut. 1991;32:905-12.
119. Oliver MG, Specian RD, Perry MA et al. Morphologic assessment of leukocyte-endothelial cell interactions in mesenteric venules subjected to ischemia and reperfusion. Inflammation. 1991;15(5):331-46.
120. Suzuki M, Inauen W, Kvietys PR et al. Superoxide mediates reperfusion-induced leukocyte-endothelial cell interactions. Am J Physiol. 1989;257:H1740-5.

121. Suzuki M, Asako H, Kubes P et al. Neutrophil-derived oxidants promote leukocyte adherence in postcapillary venules. Micro Res. 1991;42:125-38.
122. Nolte D, Hecht R, Botzlar A et al. Role of adhesion proteins during postischemic reperfusion of mouse striated muscle. Eur Surg Res. 1992;25:406-18.
123. Weselcouch EO, Grove RI, Demusz CD et al. Effect of in vivo inhibition of neutrophil adherence on skeletal muscle function during ischemia in ferrets. Am J Physiol. 1991;261:H1178-83.
124. Weibel ER, Kistler GS, Scherle WF. Practical stereological methods for morphometric cytology. J Cell Biol. 1966;30:23-38.
125. Segal S. Communication among endothelial and smooth muscle cells coordinates blood flow control during exercise. NIPS. 1992; 7:152-6.
126. Youker K, Smith CW, Anderson DC et al. Neutrophil adherence to isolated adult cardiac myocytes. Induction by cardiac lymph collected during ischemia and reperfusion. J Clin Invest. 1992;89:602-9.
127. Horgan MJ, Ge M, Gu J et al. Role of ICAM-1 in neutrophil-mediated lung vascular injury after occlusion and reperfusion. Am J Physiol. 1991;261:H1578-84.
128. Ma XL, Lefer DJ, Lefer AM, et al. Coronary endothelial and cardiac protective effects of a monoclonal antibody to intercellular adhesion molecule-1 in myocardial ischemia and reperfusion. Circulation 1992; 86:937-46.
129. Clark WM, Madden KP, Rothlein R et al. Reduction of central nervous system ischemic injury in rabbits using leukocyte adhesion antibody treatment. Stroke. 1991;22:877-83.
130. Smith CW, Entman ML, Lane CL et al. Adherence of neutrophils to canine cardiac myocytes in vitro is dependent on intercellular adhesion molecule-1. J Clin Invest. 1991;88:1216-23.
131. Bevilacqua MP, Corless C, Lo SK. Cellular and molecular mechanisms of inflammation, Vol. 2, Vascular Adhesion Molecules. San Diego: Academic Press, 1991:1-14.
132. Bevilacqua MP, Strengelin S, Gimbrone MA et al. Endothelial leukocyte adhesion molecule 1: an inducible receptor for neutrophils related to complement regulatory proteins and lectins. Science. 1989;243:1160-5.
133. Welbourn R, Goldman G, Paterson IS et al. Pathophysiology of ischaemia reperfusion injury, central role of the neutrophil. Br J Surg. 1991;78:651-5.
134. Hill J, Lindsay T, Rusche J et al. A Mac-1 antibody reduces liver and lung injury but not neutrophil sequestration after intestinal ischemia-reperfusion. Surgery. 1992;112:166-72.
135. Klausner JM, Anner H, Paterson IS et al. Lower torso ischemia-induced lung injury is leukocyte dependent. Ann Surg. 1988; 209:231-6.
136. Welbourn R, Goldman G, Hill J et al. Role of neutrophil adherence receptors (CD18) in lung permeability following lower torso ischemia. Circ Res. 1992;71:82-6.
137. Carden DL, Young JA, Granger DN. Pulmonary microvascular injury following intestinal ischemia/reperfusion: role of P-selectin. J Appl Physiol. 1993;(in press)1994.
138. Goldman G, Welbourn R, Klausner JM et al. Thromboxane mediates diapedesis after ischemia by activation of neutrophil adhesion receptors interacting with basally expressed intercellular adhesion molecule-1. Circ Res. 1991;68:1013-9.
139. Carden DL, Korthuis RJ. Neutrophil chemotactic activity generated in canine plasma by superoxide. FASEB J. 1992; 6:A2071
140. Lewis RE, Granger HJ. Diapedesis and the permeability of venous microvessels to protein macromolecules: the impact of leukotriene B4. Microvasc Res. 1988;35:27
141. Doukas J, Shepro D, Hechtman HB. Vasoactive amines directly modify endothelial cells to affect polymorphonuclear leukocyte diapedesis in vitro. Blood. 1987;69:1563-9.
142. Alexander JS, Hechtman HB, Shepro D. Phalloidin enhances endothelial barrier function and reduces inflammatory permeability in vitro. Microvasc Res. 1988;35:308-15.
143. Korthuis RJ, Carden DL, Kvietys PR et al. Phalloidin attenuates postischemic neutrophil infiltration and increased microvascular permeability. J Appl Physiol. 1991; 71:1261-9.

144. Asako H, Wolf R, Granger DN et al. Phalloidin reduces leukocyte emigration and vascular permeability in postcapillary venules. Am J Physiol. 1992; 263:H1637-42.
145. Lewis MS, Whatley RE, Cain P et al. Hydrogen peroxide stimulates the synthesis of platelet-activating factor by endothelium and induces endothelial cell-dependent neutrophil adhesion. J Clin Invest. 1988; 82:2045-55.
146. Gasic AC, McGuire G, Krater S et al. Hydrogen peroxide pretreatment of perfused canine vessels induces ICAM-1 and CD18-dependent neutrophil adherence. Circulation. 1991;84:2154-66.
147. Zimmerman BJ, Grisham MB, Granger DN. Role of hydrogen peroxide, iron, and hydroxyl radicals in ischemia/reperfusion-induced neutrophil infiltration. Physiologist (Suppl). 1988;31:A229
148. Zimmerman BJ, Grisham MB, Granger DN. Role of oxidants in ischemia/reperfusion-induced granulocyte infiltration. Am J Physiol. 1990;258:G185-90.
149. Suzuki M, Grisham MB, Granger DN. Leukocyte-endothelial cell adhesive interactions: role of xanthine oxidase-derived oxidants. J Leuk Biol. 1991;50:488-94.
150. Shingu M, Nobunaga M. Chemotactic activity generated in human serum from the fifth component of complement by hydrogen peroxide. Am J Pathol. 1984; 117:201-6.
151. Vogt W, von Zabern I, Hessp D et al. Generation of an activated form of C5 (C5b-like C5) by oxygen radicals. Immunol Lett. 1987;14:209-15.
152. Kubes P, Suzuki M, Granger DN. An endogenous modulator of leukocyte adhesion. Proc Natl Acad Sci USA. 1991;88:4651-5.
153. Kubes P, Granger DN. Nitric oxide modulates microvascular permeability. Am J Physiol. 1992;262:H611-5.
154. Kurose I, Kubes P, Wolf R et al. Inhibition of nitric oxide production: mechanisms of vascular albumin leakage. Circ Res. 1993; 73:164-71.
155. McEver RP. Novel adhesion receptors that mediate leukocyte adhesion during inflammation. Thromb Haem. 1991;65:223-8.
156. Patel KD, Zimmerman GA, Prescott SM et al. Oxygen radicals induce human endothelial cells to express GMP-140 and bind neutrophils. J Cell Biol. 1991; 112(4): 749-59.
157. Soussi B, Idstorm JP, Schersten T et al. Cytochrome C oxidase and cardiolipin alterations in response to skeletal muscle ischemia and reperfusion. Acta physiol scand. 1990;138:107-14.
158. Girotti AW. Mechanisms of lipid peroxidation. Free Radic Biol Med. 1985; 1:87-95.
159. Perry MO, Fantini G. Ischemia: Profile of an enemy. Reperfusion injury of skeletal muscle. J Vasc Surg 1987; 6:231-234.
160. Yoshioka T, Shires GT, Fantini GA. Hypothermia relieves oxidative stress in reperfused skeletal muscle following partial ischemia. J Surg Res. 1992;53:408-16.
161. Smith JK, Grisham MB, Granger DN et al. Free radical defense mechanisms and neutrophil infiltration in postischemic skeletal muscle. Am J Physiol. 1989;256:H789-93.
162. Varani J, Ginsburg I, Schuger L et al. Endothelial cell killing by neutrophils. Am J Pathol. 1989;135:435-8.
163. Vissers CM, Winterbourn CC. Oxidative damage to fibronectin. I. The effects of the neutrophil myeloperoxidase system and HOCl. Arch Biochem Biophys. 1991; 285:53-9.
164. Weiss SJ, Peppin G, Ortiz X. Oxidative auto-activation of latent collagenase by human neutrophils. Science. 1985; 222: 747-9.
165. Weiss SJ, Regiani S. Neutrophils degrade subendothelial matrices in the presence of alpha-1-proteinase inhibitor. Cooperative use of lysosomal proteinase and oxygen metabolites. J Clin Invest. 1984;73:1297-303.
166. Carden DL, Korthuis RJ. Role of neutrophilic elastase in postischemic granulocyte extravasation and microvascular dysfunction in skeletal muscle. FASEB J (Suppl). 1990;4:A1248
167. Ames A, Wright LW, Kowada JM et al. Cerebral ischemia: II. The no-reflow phenomenon. Am J Pathol. 1968;52:437-53.
168. Barrosa-Aranda J, Schmid-Schonbein GW, Zweifach BW et al. Granulocytes and the no-reflow phenomenon in irreversible hemorrhagic shock. Circ Res. 1988;63:437-47.

169. del Zappo GJ, Schmid-Schonbein GW, Mori E et al. Polymorphonuclear leukocytes occlude capillaries following middle cerebral artery occlusion and reperfusion in baboons. Stroke. 1991;22:1276-83.
170. Engler RL, Dahlgren MD, Morris DD et al. Role of leukocytes in response to acute myocardial ischemia and reflow in dogs. Am J Physiol. 1986;251:H314-23.
171. Engler RL, Dahlgren MD, Peterson MA et al. Accumulation of polymorphonuclear leukocytes during 3-h experimental myocardial ischemia. Am J Physiol. 1986; 251: H93-H100.
172. Engler RL, Schmid-Schonbein GW, Pavelec RS. Leukocyte capillary plugging in myocardial ischemia and reperfusion in the dog. Am J Pathol. 1983;111:98-111.
173. Fischer EG, Ames A, Hedley-White ET et al. Reassessment of cerebral capillary changes in acute global ischemia and their relationship to the "no-reflow phenomenon". Stroke. 1977;8:36-9.
174. Harmon JW. The significance of local vascular phenomena in production of ischemic necrosis in skeletal muscle. Am J Pathol. 1948;24:625-41.
175. Jerome SN, Smith CW, Korthuis RJ. CD18-dependent adherence reactions play an important role in the development of the no-reflow phenomenon. Am J Physiol. 1993; 263:H1637-42.
176. Kloner RA, Ganote CE, Jennings RB. The "no-reflow" phenomenon after temporary coronary occlusion in the dog. J Clin Invest. 1974;54:1496-508.
177. Korthuis RJ, Granger DN. Ischemia/reperfusion injury: role of oxygen-derived free radicals. In: Matalon S, Taylor AE, Ward PA, eds. Physiology of oxygen radicals. Bethesda, Md.: American Physiological Society, 1986:217-50.
178. Mori E, del Zappo GJ, Chambers JD, Cet al. Inhibition of polymorphonuclear leukocyte adherence suppresses no-reflow after focal cerebral ischemial in baboons. Stroke. 1992;23:712-8.
179. Schmid-Schonbein GW. Capillary plugging by granulocytes and the no-reflow phenomenon in the microcirculation. Fed Proc. 1987;46:2397-401.
180. Strock PE, Majno GM. Microvascular changes in acutely ischemic rat muscle. Surg Gynecol Obstet. 1969;129:1213-23.
181. Quiñones-Baldrich WJ, Chervu A, Hernandez JJ et al. Skeletal muscle function after ischemia: "no-reflow" versus reperfusion injury. J Surg Res. 1991;51:5-12.
182. Messina LM. In vivo assessment of acute microvascular injury after reperfusion of ischemic tibialis anterior muscle of hamster. J Surg Res. 1990;48:615-21.
183. Harman JW. The significance of local vascular phenomenon in production of ischemic necrosis in skeletal muscle. Am J Pathol. 1941;24:625-41.
184. Bagge U, Amundson B, Lauritzen C. White blood cell deformability and plugging of skeletal muscle capillaries in hemorrhagic shock. Acta Physiol Scand. 1980; 108: 159-63.
185. Jerome SN, Dore M, Paulson JC et al. P-selectin and ICAM-1 dependent adherence reactions: role in the genesis of postischemic no-reflow. Am J Physiol (Heart Circ Physiol). 1994; 266:H1316-H1321.
186. Granger DN, Benoit JN, Suzuki M et al. Leukocyte adherence to venular endothelium during ischemia-reperfusion. Am J Physiol. 1989;257:G683-8.
187. Menger MD, Steiner D, Messmer K. Microvascular ischemia-reperfusion injury in striated muscle: significance of "no-reflow". Am J Physiol. 1992;263:H1892-900.
188. Hansell P, Borgstrom P, Arfors KE. Pressure-related capillary leukostasis following ischemia-reperfusion and hemorrhagic shock. Am J Physiol. 1993;265:H381-8.
189. Mazzoni MC, Borgstrom P, Intaglietta M et al. Lumenal narrowing and endothelial cell swelling in skeletal muscle capillaries during hemorrhagic shock. Circ Shock. 1989;29:27-39.
190. Gidlof A, Lewis DH, Hammersen F. The effect of prolonged total ischemia on the ultrastructure of human skeletal muscle capillaries. A morphometric analysis. Int J Microcirc Clin Exp. 1982;7:67-86.
191. Mazzoni MC, Borgstrom P, Intaglietta M et al. Capillary narrowing in hemorrhagic shock is reduced by hyperosmotic saline-dextran reinfusion. Circ Shock. 1990; 31:407-18.

192. Jerome SN, Akimitsu T, Korthuis RJ. Leukocyte adhesion, edema, and the development of postischemic capillary no-reflow. Am J Physiol. in press, 1994.
193. Mazzoni MC, Intaglietta M, Crogue Jr. et al. Amiloride sensitive Na^+ pathways in capillary endothelial cell swelling during hemorrhagic shock. J Appl Physiol. 1992; 73:1467-73.
194. House SD, Lipowsky HH. Leukocyte-endothelium adhesion: microhemodynamics in mesentery of the cat. Microvasc Res. 1987;34:363-79.
195. Hammersen F, Barker JH, Gidlof A et al. The ultrastructure of the microvessels and their contents following ischemia and reperfusion. Prog Appl Microcirc. 1989; 13:1-26.
196. Menger MD, Sack FU, Barker JH et al. Quantitative analysis of microcirculatory disorders after prolonged ischemia in skeletal muscle. Therapeutic effects of prophylactic isovolemic hemodilution. Res Exp Med. 1988;188:151-66.

CHAPTER 3

Role of Neutrophils

Richard E. Kirschner

Gary A. Fantini

The accumulation of inflammatory cells in postischemic skeletal muscle has been recognized for decades.[1] Until recently, however, it was widely believed that leukocyte infiltration represented the normal host response to ischemic injury and that the function of extravasated polymorphonuclear leukocytes (PMNs) was primarily one of healing and repair. There is now substantial evidence implicating neutrophils in the genesis of reperfusion-induced microvascular dysfunction and myocyte necrosis, suggesting that leukosequestration is a cause, rather than simply an effect, of postischemic injury in skeletal muscle.[2-5] Numerous studies have indicated that reperfusion of ischemic skeletal muscle results in the elaboration of proinflammatory mediators responsible for the activation of circulating PMNs and the promotion of neutrophil chemotaxis and adherence to the microvascular endothelium.[6-10] The subsequent release of reactive oxygen metabolites and cytotoxic lysosomal proteases by adherent or extravasated neutrophils may degrade components of the extracellular matrix and peroxidize membrane phospholipids, thus promoting injury to the postischemic microvasculature and skeletal myocytes. In addition, there is considerable experimental evidence indicating that leukocyte plugging of the microvasculature may contribute to ongoing ischemic injury through the development of the no-reflow phenomenon.[11-13]

NEUTROPHIL INFILTRATION

The importance of neutrophil infiltration in the generation of reperfusion injury was first demonstrated in studies of the postischemic myocardium.[14,15] Histologically, myocardial infarction resembles an acute inflammatory process characterized by an extensive early accumulation of extravasated neutrophils.[16] Romson et al[14] reported that depletion of circulating PMNs in dogs by the administration of specific rabbit antiserum to dog neutrophils was associated with a significant reduction in the extent of neutrophil

Ischemia-Reperfusion Injury of Skeletal Muscle, edited by Gary A. Fantini, M.D.;

infiltration and myocardial infarct size following temporary left circumflex coronary artery occlusion. A subsequent report by Mullane et al[15] confirmed that depletion of circulating neutrophils significantly reduced myocardial leukosequestration and diminished infarct size. Based upon these observations, the investigators concluded that neutrophil accumulation in response to myocardial ischemia was an important determinant of the extent of postischemic myocardial injury.

As in the myocardium, restoration of blood flow to ischemic skeletal muscle is associated with massive leukocyte infiltration.[1,17] The role of PMNs in the pathogenesis of the vascular injury associated with skeletal muscle ischemia-reperfusion was first demonstrated in 1988 by Korthuis and coworkers,[3] who reported that leukocyte depletion using Leukopak filters attenuated reperfusion-induced increases in microvascular permeability and skeletal muscle vascular resistance. Similar results were later observed in animals rendered neutropenic by the administration of specific polyclonal antineutrophil serum.[18] Other investigators have reported that neutrophil depletion with chemotherapeutic agents attenuates the rise in lymph flow and lymph protein clearance in canine hindlimbs reperfused after 2 hours of tourniquet ischemia,[10] again indicating that leukocytes play a critical role in the pathogenesis of microvascular injury following skeletal muscle ischemia-reperfusion.

In addition to contributing to alterations in microvascular permeability, neutrophil infiltration may also exacerbate skeletal myocyte injury following ischemia. High-grade partial hindlimb ischemia in the rat is associated with significant membrane dysfunction, as indicated by a depolarization of the resting skeletal muscle transmembrane potential difference (E_m).[19,20] This depolarization of the cell membrane persists following the restoration of blood flow despite the complete repletion of intracellular high energy phosphates, signifying that membrane dysfunction is an ongoing process during reperfusion. In contrast, repolarization of the resting E_m toward preischemic values is observed during hindlimb reperfusion in neutropenic animals.[4] In other studies, Walden et al[2] have examined the effect of vinblastine-induced neutropenia on skeletal muscle function in postischemic rat hindlimbs. After 3 hours of unilateral tourniquet ischemia and 4 hours of reperfusion, electrically stimulated gastrocnemius muscles of untreated animals were found to generate approximately 14% of the force of contraction compared with the nonischemic contralateral hindlimb. In rats rendered neutropenic by pretreatment with vinblastine, recovered muscle function was significantly greater than that of untreated animals, with the force of contraction reaching 40% of that measured in the nonischemic control limbs. Further evidence for the role of neutrophils in the genesis of skeletal myocyte injury following ischemia-reperfusion is provided by experiments in which radiation-induced neutropenia substantially reduces the extent of infarction in gastrocnemius muscles, as quantitated by the reduction of triphenyltetrazolium chloride, in rats subjected to 3 hours of unilateral hindlimb tourniquet ischemia and 1 hour of reperfusion.[5] Together, these studies implicate PMNs as important mediators of myocyte injury following skeletal muscle ischemia-reperfusion.

THE ROLE OF NEUTROPHIL ADHESION

An increase in the capacity for neutrophil adherence to the endothelial lining of the microvasculature is an important factor in the development of microvascular and skeletal myocyte injury following ischemia-reperfusion. Recent evidence indicates that neutrophil binding to the microvascular endothelium is a critical step in the sequence of events leading to neutrophil activation and to the release of cytotoxic proteases and oxygen-derived free radicals.[21,22] Moreover, the establishment of specific adhesive interactions between neutrophils and vascular endothelial cells appears to be an essential prerequisite to neutrophil diapedesis and the accumulation of activated granulocytes within postischemic tissues.[18,23]

Under normal conditions, the endothelium of postcapillary venules has little capacity to recognize and bind circulating leukocytes. As a result, neutrophils travel through the microcirculation at a velocity determined predominantly by the dynamics of blood flow.[24] In response to local inflammatory signals, however, circulating neutrophils may become displaced toward the vessel wall, an event referred to as margination. Transient adhesive interactions between marginated neutrophils and the microvascular endothelium produces leukocyte rolling, a process whereby neutrophils travel through the microcirculation at a substantially reduced velocity as compared with the rate of blood flow. Observations of the deformation of rolling leukocytes by the forces of blood flow in postcapillary venules indicate that these cells are in fact engaged in a process of transient adhesion and detachment.[25,26] The subsequent development of stronger adhesive forces between marginated leukocytes and vascular endothelial cells is indicated by the spreading of stationary neutrophils along the microvascular endothelium. This firm attachment inevitably precedes the emigration of activated neutrophils from the intravascular space (Fig. 3.1; see frontispiece opposite title page).

The adherence of neutrophils to the microvascular endothelium in postischemic skeletal muscle and other tissues has been demonstrated utilizing the technique of intravital microscopy. Erlansson et al[27] studied leukocyte behavior in the hamster cheek pouch before and after a 30 minute ischemic insult. Leukocytes adhering to the endothelium of segments of postcapillary venules were examined at high power magnification, and those adhering for at least 5 seconds were counted. When these values, divided by the vessel volumes, were utilized as an index of leukocyte adhesion, the number of adherent leukocytes observed during the preischemic control period varied between zero and 0.03 cells/pl. After reperfusion, the index increased dramatically, reaching a peak value between 0.23 and 0.83 cells/pl in just 5 to 15 minutes ($P < 0.01$). Intravital microscopic observation of leukocyte adherence to the postischemic microvascular endothelium in striated muscle has been reported by Lehr and coworkers.[28,29] Under normal conditions, the majority of leukocytes were observed to pass through the postcapillary venules of hamster dorsal skinfold preparations without interacting with the endothelium. Reperfusion after four hours of pressure-induced ischemia elicited the rolling of labelled neutrophils along the endothelium and the subsequent firm adhesion of leukocytes to postcapillary venules, with maximal leukocyte adhesion apparent 30 minutes following the restoration of blood flow. Postischemic adhesion of leukocytes to the striated muscle microvasculature was associated with a significant loss of endothelial integrity, as measured by an increase in macromolecular leakage into the interstitial space. In similar studies, Goldberg et al[30] demonstrated that ischemia-reperfusion of rat cremaster results in enhanced rolling of labelled leukocytes along the microvascular endothelium.[30] Reperfusion following a three hour period of ischemia increased the number of neutrophils rolling along the endothelial lining of postcapillary venules from 9.5 per minute in control animals to 24.1 per minute in those with cremasters subjected to ischemia-reperfusion ($P < 0.05$).

Within the past decade, experiments have demonstrated that the complex sequence of adhesive interactions between leukocytes and the microvascular endothelium is controlled by the expression of specific adhesion receptors on surface of both cell types.[31,32] The adhesion molecules that have been characterized to date are currently classified into three structurally distinct groups: the selectin family, the integrin family, and the immunoglobulin superfamily (Table 3.1).

The Selectin Family

Leukocyte rolling is mediated by the selectin family of transmembrane glycoprotein adhesion molecules.[25] The selectins are comprised of three structurally similar monomeric glycoproteins each containing an extracellular N-terminal domain that is homologous to the calcium-dependent lectins (Fig. 3.2). This lectin-like domain is followed by a domain homologous to epidermal growth factor, a variable number of short consensus repeat sequences characteristic of proteins interacting with complement components, a transmembrane domain, and finally a short

Table 3.1. Adhesion molecules involved in leukocyte-endothelial interactions

Receptor Name	Other Names	Cell Distribution	Ligand
Selectins			
L-selectin	gp90Mel, Mel-14, LAM-1, LECCAM-1	Neutrophils Lymphocytes	Sialylated ?
E-selectin	ELAM-1	Endothelial cells	Sialylated Lewis X, ?
P-selectin	GMP-140, PADGEM, CD62	Endothelial cells Platelets	Sialylated Lewis X, ?
β_2-Integrins			
LFA-1 (CD11a/CD18)		Neutrophils, monocytes, macrophages, lymphocytes, NK cells	ICAM-1, ICAM-2
Mac-1 (CD11b/CD18)	CR3, Mo-1	Neutrophils, monocytes, macrophages, NK cells	ICAM-1, ?
gp150,95 (CD11c/CD18)	CR4	Neutrophils, monocytes, macrophages, lymphocytes, NK cells	?
Ig Family			
ICAM-1	CD54	Endothelial cells	LFA-1, Mac-1
ICAM-2		Endothelial cells	LFA-1

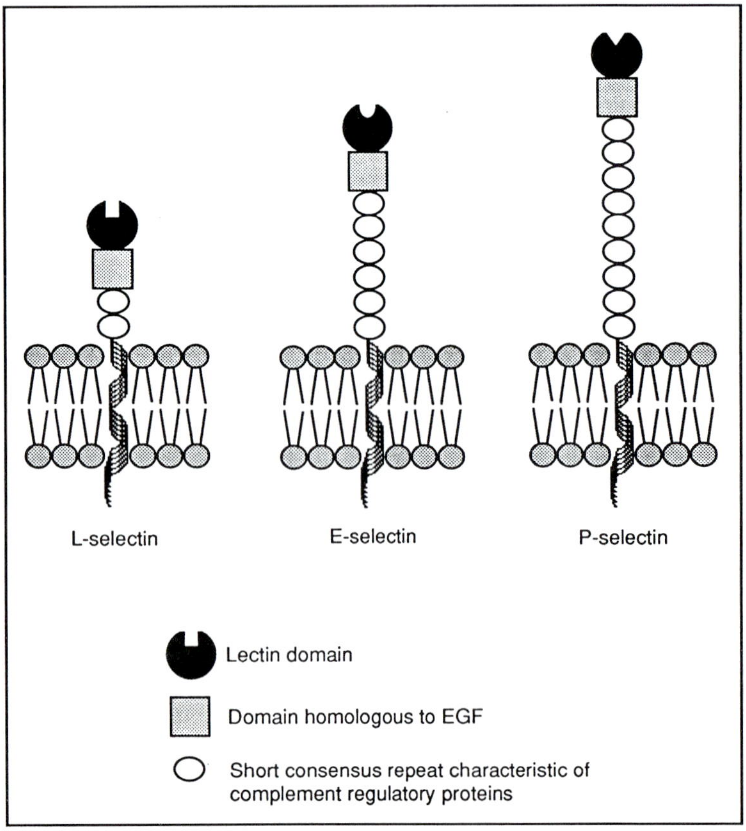

Fig. 3.2. Structure of the selectins.

intracellular C-terminal domain.[32,33] The smallest member of the selectin family is L-selectin (previously designated LAM-1, LECCAM-1, gp90Mel, Mel-14 antigen), which is present on the surface of neutrophils, monocytes, and lymphocytes and which binds to both constitutive and inducible carbohydrate ligands on the endothelial cell surface. Blocking L-selectin with polyclonal[34] and monoclonal[35,36] antibodies or with recombinant soluble L-selectin[34] impairs the ability of neutrophils to roll along cultured endothelial monolayers and along the endothelial surface of mesenteric venules, indicating that L-selectin-mediated interactions are essential to leukocyte rolling. L-selectin is rapidly shed from the cell surface following leukocyte activation,[37] an event which is associated with the inability of these cells to migrate to inflammatory sites. This down-regulation of cell surface adhesion molecules may allow activated neutrophils to detach from the endothelial surface and extravasate into the extravascular space. The migration of neutrophils into inflammatory sites is significantly attenuated by L-selectin blockade,[37,38] thus underscoring the importance of early L-selectin-dependent interactions between neutrophils and the microvascular endothelium in the process of leukosequestration.

Leukocyte rolling in vitro[39] and in vivo[40] has also been demonstrated to involve the interaction between neutrophils and the endothelial cell surface glycoprotein P-selectin. The largest of the selectin family, P-selectin (CD62, PADGEM, GMP-140) is constitutively associated with the α-granules of platelets[41] and the cytoplasmic Weibel-Palade bodies of endothelial cells[42,43] and is rapidly translocated to the endothelial cell surface following stimulation by histamine or thrombin.[43] P-selectin expression may also be induced by exposure of the vascular endothelium to oxygen-derived free radicals,[44] the complement complex C5b-9,[45] and tumor necrosis factor-α (TNF-α).[46] The extracellular N-terminal domain of the molecule binds to an as yet unidentified carbohydrate ligand on the surface of circulating leukocytes to promote early leukocyte adhesion during the inflammatory response.[47,48]

The third member of the selectin family is E-selectin (endothelial leukocyte adhesion molecule-1, ELAM-1), a glycoprotein which is found exclusively on the surface of endothelial cells. Although not expressed on resting endothelium, E-selectin is markedly upregulated in response to the cytokines interleukin-1 (IL-1) and TNF-α and to lipopolysaccharide.[49] The apparent carbohydrate ligand for E-selectin contains the sialylated Lewis^x determinant present on the surface of circulating leukocytes.[50-53] Like the other selectins, E-selectin has been shown play an important role in leukocyte adhesion to the microvascular endothelium.[49,54]

The Integrin Family

The integrins comprise a vast number of cell surface adhesion molecules that mediate intercellular recognition and cellular binding to the extracellular matrix.[55] Members of this group of transmembrane receptors are heterodimers composed of α and β subunits.[56] The β_2 subfamily of integrins, known as the leukocyte integrins or LeuCAMs, is expressed exclusively by leukocytes and is comprised of three related heterodimers, each composed of a common β subunit (CD18) which is complexed with unique α subunit (CD11a, CD11b, or CD11c).[57,58]

LFA-1 (lymphocyte function-associated antigen-1; CD11a/CD18) is expressed on all leukocytes and mediates the attachment of unstimulated neutrophils to the vascular endothelium through a specific interaction with intercellular adhesion molecule-1 (ICAM-1) and intercellular adhesion molecule-2 (ICAM-2), which are described below.[31]

Mac-1 (CD11b/CDF18), previously referred to as complement receptor 3 (CR3) because of its ability to bind the activated complement product C3b, is constitutively expressed on the surface of neutrophils, monocytes, and macrophages and binds to endothelial ICAM-1 at a site distinct from the LFA-1 binding domain.[59,60] In addition to that expressed on the cell surface, an intracellular pool of preformed Mac-1 is stored within neutrophil specific granules. Translocation of Mac-1 from intracellular pools to the neutrophil surface via granule fusion with the plasma membrane is stimulated by chemotaxins, including complement fragment C5a, leukotriene B_4 (LTB_4), interleukin-8, and platelet-activating factor (PAF).[61] In contrast to the adhesive interactions between LFA-1 and ICAM-1, which can be demonstrated with both stimulated and unstimulated cells, Mac-1-mediated adherence to endothelial ICAM-1 appears to require chemotactic activation.[59,62]

The third of the leukocyte integrins, designated gp150,95 (CD11c/CD18), is sometimes referred to as complement receptor 4 (CR4) because it may also bind C3b. The gp150,95 molecule has been identified on the surface of neutrophils and monocytes and is believed to participate in monocyte adhesion. Its endothelial ligand is as yet unknown.[63]

The Immunoglobulin Superfamily

The leukocyte integrins bind to endothelial surface receptors belonging to the immunoglobulin supergene family, the members of which share one or more immunoglobulin domains.[64] The two members of this family currently known to participate in neutrophil binding are ICAM-1 (CD54) and ICAM-2. ICAM-1 is the primary physiologic ligand for LFA-1[65,66] and has been shown to interact with Mac-1 through a separate binding domain.[59] ICAM-1 expression, like that of E-selectin, is markedly enhanced by endothelial exposure to cytokines or lipopolysaccharide.[67] Unlike E-selectin, however, ICAM-1 is constitutively expressed at low levels on the surface of resting endothelial cells. Unstimulated endothelium also expresses ICAM-2, which serves as a ligand for LFA-1.[68]

Although the exact mechanisms responsible for enhanced neutrophil-endothelial interaction following ischemia-reperfusion remain uncertain, leukocyte binding appears to follow an orderly sequence of events mediated by adhesion molecules on the surface of both cell types (Fig. 3.3). In the resting state, leukocyte integrins remain in an inactive conformation, and the endothelium expresses ICAM-2 and small amounts of ICAM-1. P-selectin, E-selectin, and the ligand for L-selectin are not expressed on the unstimulated endothelial surface. Thus, although circulating neutrophils display L-selectin and the carbohydrate ligands for P- and E-selectin, leukocyte adhesion to the endothelium can not occur. Following ischemia-reperfusion, P-selectin is rapidly expressed on the endothelial surface, resulting in early leukocyte adhesion and neutrophil rolling.[39,69] While numerous proinflammatory mediators have been demonstrated to mediate the translocation of P-selectin from the cytoplasmic Weibel-Palade bodies of endothelial cells to the plasma membrane in vitro,[44-46,70] the exact nature of the physiologically relevant effectors of P-selectin upregulation following ischemia remains to be elucidated. Unlike integrin-mediated adhesion to the endothelium, neutrophil binding to P-selectin does not require PMN activation. Recent experiments indicate that neutrophils adherent to endothelial cells expressing P-selectin become polarized in shape, activate surface β_2-integrins, and are primed for enhanced degranulation upon subsequent stimulation with chemotactic factors.[21] This activation of leukocytes bound to P-selectin may be inhibited by blocking PAF receptors on the neutrophil surface.[21] Thus, P-selectin and PAF expressed by activated endothelial cells appear to act cooperatively in a cascade of PMN adhesion and activation wherein P-selectin tethers the neutrophil to the endothelial surface and thus facilitates its interaction with the signaling molecule, PAF. The juxtacrine stimulation of PMNs by endothelium-derived chemotactic factors initiates a second phase of neutrophil adhesion that is characterized by the shedding of L-selectin from the neutrophil surface, the enhancement of Mac-1 binding to endothelial ICAM-1, and the arrest of PMN rolling.[31,39] Although chemotactic stimulation has been shown to result in quantitative upregulation of Mac-1 on the neutrophil surface, increased Mac-1 expression is neither necessary nor sufficient for the increased adhesiveness of activated leukocytes.[71] Rather, the enhancement of integrin-mediated adhesion following neutrophil activation appears to result from a qualitative change in the Mac-1 expressed on the leukocyte

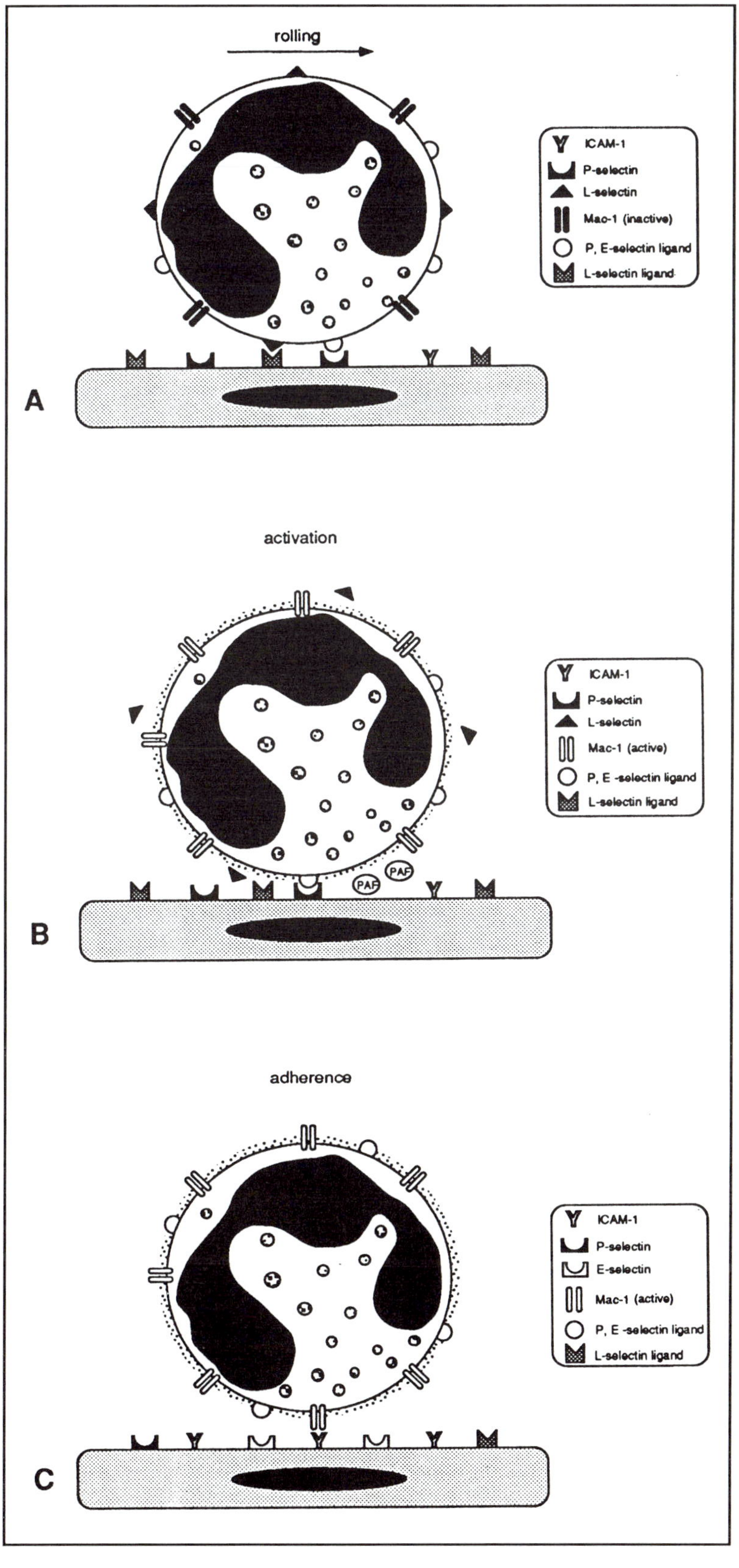

Fig. 3.3. Sequence of receptor-mediated events in neutrophil adhesion. In the resting state, endothelial cells express ICAM-2 (not shown) and low levels of ICAM-1. Circulating leukocytes express L-selectin and the carbohydrate ligands for P- and E-selectin (ELAM-1). The leukocyte integrin Mac-1 is present on the neutrophil surface (as is LFA-1, not shown), but is in an inactive state and hence does not bind to its endothelial ligand, ICAM-1. (A) Following stimulation by proinflammatory mediators such as histamine or thrombin, the endothelium expresses P-selectin and the ligand for L-selectin, resulting in leukocyte rolling. Mac-1 remains inactive. (B) Juxtacrine stimulation of rolling leukocytes by endothelium-derived platelet-activating factor (PAF) and perhaps other chemoattractants activate Mac-1 on the neutrophil surface, thus allowing its interaction with ICAM-1. Additional Mac-1 may be translocated to the cell membrane from granular stores. Upon neutrophil activation, L-selectin is rapidly shed from the cell surface. (C) Cytokine stimulation results in the upregulation of ICAM-1 on the endothelial surface and the expression of E-selectin. Neutrophil binding to these receptors results in firm leukocyte adhesion to the endothelium and is prerequisite to subsequent diapedesis.

surface, the exact biochemical nature of which remains uncertain.[24] Recent data indicate that enhanced leukocyte adherence to the endothelium may also result from the "unmasking" of Mac-1 binding sites on ICAM-1 in the presence of activated neutrophils.[60] Ultimately, cytokine stimulation promotes the expression of E-selectin and the upregulation of ICAM-1 on the endothelial surface. Experimental evidence suggests that binding to both E-selectin and ICAM-1 is prerequisite to the subsequent migration of activated neutrophils across the endothelium and into the extravascular space.[72]

The important role of adhesion molecules in promoting leukocyte adherence to postischemic endothelium has been established in a number of studies by utilizing monoclonal antibodies directed against specific cell surface glycoproteins. In vivo neutralization of P-selectin has been documented to significantly reduce leukocyte infiltration and reperfusion injury in the feline myocardium[73] and in the rabbit ear,[69] indicating that the early interaction between P-selectin and its ligand on the neutrophil surface is of critical importance in the pathogenesis of ischemia-reperfusion injury. Moreover, the administration of monoclonal antibodies directed against the common β-subunit (CD18) of the leukocyte integrins has been shown to dramatically reduce neutrophil adherence to feline mesenteric venules subjected to ischemia-reperfusion[74] and to cultured human umbilical vein endothelial cell (HUVEC) monolayers exposed to anoxia-reoxygenation (A/R).[75] Yoshida et al[75] have observed that neutrophil adhesion to HUVEC monolayers reoxygenated after 30 minutes of anoxia increased 2.7-fold as compared with that measured in normoxic controls. The postanoxic increase in in neutrophil adherence was significantly attenuated by monoclonal antibodies directed against the unique integrin α-subunits, CD11a and CD11b, and the common β-subunit, CD18, suggesting that the enhanced leukocyte adherence induced by anoxia-reoxygenation involves both LFA-1 (CD11a/CD18) and Mac-1(CD11b/CD18). Monoclonal antibodies directed against ICAM-1, the endothelial ligand for both LFA-1 and Mac-1, were similarly observed to attenuate postanoxic leukocyte adhesion. Flow cytometric studies demonstrated that incubation of neutrophils with supernatants of A/R-conditioned HUVEC monolayers elicited an increase in the cell surface expression of CD11b and CD18, but not CD11a. In addition, supernatants from HUVEC subjected to A/R promoted neutrophil adherence to naive HUVEC, an effect that was diminished by PAF receptor blockade. Thus, A/R-induced adherence of neutrophils to endothelial monolayers appears to involve the specific interactions between integrins expressed on the neutrophil surface and ICAM-1 on the vascular endothelium that appear to be mediated by PAF.

The importance of integrin-mediated interactions in the pathogenesis of ischemia-reperfusion injury in skeletal muscle has been recently elucidated utilizing monoclonal antibodies directed against the functional epitopes of the leukocyte integrins. Pretreatment with the anti-CD18 monoclonal antibody MoAb 60.3 inhibits both neutrophil adhesion and transendothelial migration in skeletal muscle following stimulation with LTB_4 and zymosan-activated serum.[76] Leukocyte rolling along the skeletal muscle microvascular endothelium is not affected by MoAb 60.3, supporting the notion that the early interaction between neutrophils and the vascular endothelium is a CD18-independent process, whereas later neutrophil adhesion involves CD18-mediated interactions. Carden and coworkers[18] have reported that administration of the anti-CD18 monoclonal antibody IB_4 protects against increases in microvascular resistance and permeability in isolated canine gracilis preparations subjected to ischemia-reperfusion and is as effective in this regard as neutrophil depletion with specific polyclonal antineutrophil serum. Moreover, pretreatment with IB_4 dramatically reduces postischemic gracilis muscle myeloperoxidase (MPO) activity, a measure of tissue neutrophil content, to levels indistinguishable from those observed in neutrophil depleted animals. Thus, the interaction between leukocyte integrins and the vascular endothelium is of critical importance in promoting leukosequestration and the genesis of microvascular dysfunction in reperfused skeletal muscle.

Prevention of integrin-mediated neutrophil adherence has also been shown to prevent ischemia-induced skeletal muscle contractile dysfunction. Weselcouch et al[77] produced hindlimb ischemia in ferrets by partial occlusion of the abdominal aorta, resulting in a 60% reduction of resting blood flow in the muscles of the distal hindlimb. The ischemic muscles were then stimulated electrically to contract isometrically and the force of contraction was recorded. In nonischemic hindlimbs, the contractile force peaked soon after the onset of stimulation and gradually declined to approximately 80% of peak force after 20 minutes. During ischemia, peak force of contraction was reduced by 25 ± 7% and within 5 minutes the force declined to approximately 20% of the original peak. In animals treated with MoAb 60.3, peak force was 94 ± 3% of that measured during the corresponding normal flow period, a significant protective effect compared with untreated controls ($P < 0.05$). In addition, the plateau of contractile forced reached in the later stages of electrical stimulation was also higher in the MoAb 60.3 group than in the control group. These findings indicate that integrin-mediated interactions between neutrophils and the vascular endothelium may contribute to skeletal muscle dysfunction induced by critical ischemia.

Although it appears that the leukocyte integrins play an important role in modulating neutrophil adhesion in postischemic skeletal muscle, there is little data available on the function of other adhesion molecules in the pathogenesis of leukocyte-mediated skeletal muscle reperfusion injury. As discussed above, recent studies have implicated P-selectin in the important early interaction between neutrophils and endothelial cells in the postischemic myocardium[73] and in the rabbit ear.[69] The importance of P-selectin in mediating leukocyte adhesion in postischemic skeletal muscle remains to be examined. A role for ICAM-1 is suggested by the previously described studies of its neutrophil ligands, LFA-1 (CD11a/CD18) and Mac-1 (CD11b/CD18); however, there is heretofore little direct evidence of its involvement in neutrophil adhesion in postischemic skeletal muscle. Recent studies have demonstrated an increase in endothelial ICAM-1 synthesis and surface expression following myocardial ischemia-reperfusion.[78] The administration of anti-ICAM-1 monoclonal antibodies substantially diminishes neutrophil infiltration in postischemic myocardium and significantly reduces myocardial infarct size.[79] In the lung, pulmonary artery occlusion induces vascular endothelial ICAM-1 upregulation, and pretreatment with anti-ICAM-1 monoclonal antibodies prevents leukosequestration, as measured by lung MPO activity, and the development of pulmonary edema following reperfusion.[80]

Several recent studies have been aimed at elucidating the role of endogenous antiadhesive metabolites in modulating neutrophil interactions with the endothelium following ischemia-reperfusion. Nitric oxide (NO), formerly termed endothelium-derived relaxing factor (EDRF), is a biologically active compound that is constitutively synthesized by endothelial NO synthase from the guanidino group of the amino acid L-arginine.[81] Initially recognized for its ability to induce vascular smooth muscle relaxation, NO has recently received considerable attention because of its apparent ability to inhibit the adhesive interaction between neutrophils and the microvascular endothelium,[82] a property that appears to result from its ability to inactivate the adhesion-promoting superoxide anion.[83] Inhibition of endothelial NO synthesis using analogues of L-arginine, such as N^G-nitro-L-arginine-methyl ester (L-NAME) and N^G-monomethyl-L-arginine (L-NMMA), has been observed to promote leukocyte adherence to mesenteric venules.[82] Studies in the feline myocardium indicate that ischemia impairs the ability of the endothelium to synthesize NO[84,85] and that the administration of exogenous L-arginine,[86] organic NO donors,[87,88] or authentic NO gas[89] prevents postischemic leukosequestration and reduces myocardial infarct size. The role of nitric oxide in the pathogenesis of skeletal muscle ischemia-reperfusion injury remains to be fully explored.

Another protective substance produced by the endothelium is the purine nucleoside adenosine. Adenosine has been shown to inhibit granulocyte function and may also inhibit neutrophil-endothelial interactions under normal physiologic conditions.[90-94] Cronstein et al[92,94] have demonstrated that adenosine, through a specific interaction with cell surface adenosine A_2 receptors, inhibits the elaboration of reactive oxygen metabolites by FMLP-stimulated neutrophils and prevents the adhesion of activated PMNs to the endothelium. Exogenous administration of adenosine inhibits ischemia-reperfusion-induced leukocyte adherence in the small bowel mesentery[95] and in skeletal muscle.[96] Since adenosine is also produced by neutrophils, it has been proposed that it might serve as an autocrine regulator of PMN function following ischemia-reperfusion. Activation of PMNs significantly attenuates the activity of their own membrane-bound 5'-nucleotidase and thereby reduces adenosine release.[97] These findings suggest alterations in adenosine metabolism may play an important role in the pathophysiology of ischemia-reperfusion.

MECHANISMS OF NEUTROPHIL-MEDIATED INJURY

Several mechanisms exist by which neutrophils may induce injury in skeletal muscle subjected to ischemia-reperfusion (Table 3.2). Activated neutrophils may produce and release reactive oxygen metabolites and toxic lysosomal enzymes into their local microenvironment, resulting in the degradation of endothelial cells, skeletal myocytes, and components of the extracellular matrix. Ordinarily, these toxins are utilized by neutrophils to protect the host against invading pathogens. As in other inflammatory conditions, however, neutrophils may be activated by ischemia-reperfusion to direct these defenses extracellularly into nearby host tissues rather than into intracellular phagocytic vacuoles. As neutrophils are activated by the appropriate stimuli in postischemic tissues, membrane-bound NADPH oxidase systems may suddenly generate large quantities of superoxide anion ($O_2^-\bullet$) and intracellular granules may fuse with the plasma membrane, discharging their toxins into the extracellular space. Potent oxidants may also be generated by the myeloperoxidase system of activated neutrophils. The result is a direct attack by neutrophil toxins on the surrounding host tissues. There is considerable experimental evidence implicating oxygen-derived free radicals as important mediators of the tissue injury produced by neutrophils in postischemic skeletal muscle; however, substantially less information is available on the role of neutrophil-derived myeloperoxidase products and proteolytic enzymes in the generation of ischemia-reperfusion injury. Several recent studies have indicated that the neutrophil may in fact utilize both the NADPH oxidase system and the granule constituents in a cooperative manner, with the concerted action of neutrophil-derived oxidants and proteinases allowing the cell to maximize its destructive potential.[98]

THE NADPH OXIDASE SYSTEM AND REACTIVE OXYGEN METABOLITES

Because of their ability to participate in reactions involving the transfer of single electrons, oxygen-derived free radicals are highly reactive molecules that are capable of inducing considerable damage in biologic systems, including DNA base modification and strand scission, protein fragmentation and cross-linking, cell matrix degradation, and lipid peroxidation.[99-101] Under normal conditions, reactive oxygen metabolites are effectively

Table 3.2. Mechanisms of neutrophil-mediated tissue injury in postischemic skeletal muscle

Reactive Oxygen Metabolites
superoxide anion
hydrogen peroxide
(?) hydroxyl radical
Myeloperoxidase products
hypochlorous acid
Lysosomal enzymes
elastase
collagenase
gelatinase
No-reflow phenomenon

neutralized by endogenous free radical scavengers such as superoxide dismutase (SOD), which catalyzes the dismutation of $O_2^{-\bullet}$ into hydrogen peroxide (H_2O_2), and glutathione peroxidase which degrades intracellular H_2O_2 into water.[101,102] Other naturally occurring antioxidants, such as α-tocopherol (vitamin E), may act at the level of the cell membrane to directly terminate the chain reactions of lipid peroxidation.[103] Following ischemia, however, the unleashing of free radical cascades may overwhelm these endogenous defenses and damage reoxygenated tissues.

Experimental evidence suggests that the peroxidation of membrane lipids is a critical component of free radical-mediated cell injury following ischemia-reperfusion. Oxygen-derived free radicals can initiate reactions of lipid peroxidation by abstracting hydrogen atoms ($H^\bullet$) from unsaturated fatty acid side chains (LH), resulting in the formation of carbon-centered lipid radicals ($L^\bullet$). Double bond rearrangement within these unstable intermediates produces conjugated dienes, which may react with molecular oxygen to yield peroxy radicals.[99] These latter may themselves remove hydrogen atoms from nearby fatty acid side chains, thereby propagating the chain reaction of oxidative membrane injury (Fig. 3.4). Unless terminated

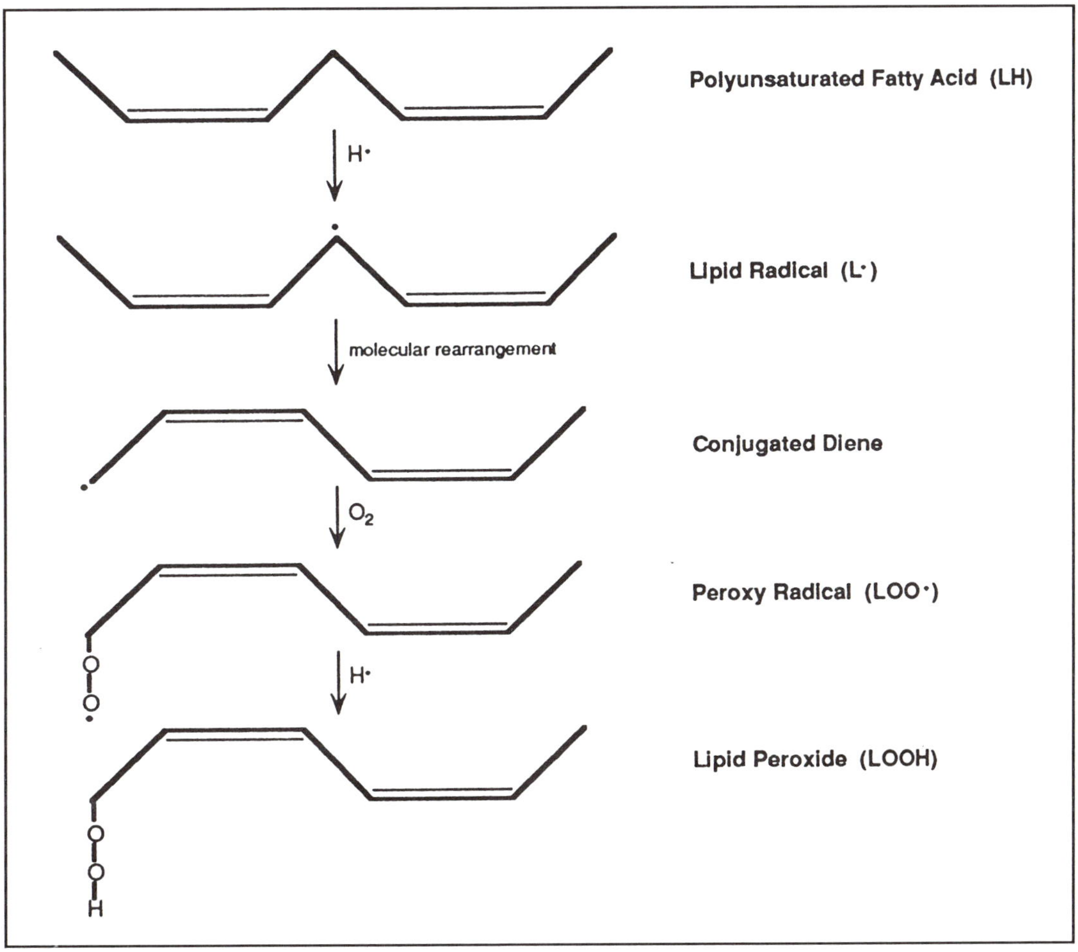

Fig. 3.4. Sequence of free radical-mediated lipid peroxidation. Abstraction of a hydrogen atom from an unsaturated fatty acid side chain forms a carbon centered lipid radical. This unstable molecule then undergoes molecular rearrangement to form a conjugated diene, which may then react with molecular oxygen to yield a peroxy radical. The chain reaction is propagated as the peroxy radical attacks an adjacent unsaturated fatty acid side chain, abstracting a hydrogen atom and leaving behind a lipid radical as it is converted into a stable lipid peroxide.

by appropriate antioxidants, lipid peroxidation results in decreased membrane fluidity, disruption of membrane-bound enzyme and transport systems, alterations in cell permeability, and cell death.[99,100]

Following exposure to an activating stimulus, polymorphonuclear leukocytes undergo a "respiratory burst", which is characterized by an increase in oxygen consumption and the generation of oxygen-derived free radicals.[104,105] The majority of oxygen consumed by neutrophils during the respiratory burst can be accounted for by the generation of $O_2^{-\bullet}$ by the NADPH oxidase system, an enzyme complex associated with the cell membrane.[106,107] While inactive in resting leukocytes, the NADPH oxidase system is rapidly activated upon neutrophil stimulation, resulting in the transfer of electrons from cytosolic NADPH to extracellular oxygen and the concomitant generation of $O_2^{-\bullet}$:[108,109]

$$2O_2 + NADPH \xrightarrow{\text{NADPH oxidase}} 2\,O_2^{-\bullet} + NADP^+ + H^+$$

Two molecules of $O_2^{-\bullet}$ may then react with one another, forming H_2O_2 and molecular oxygen in a reaction catalyzed by superoxide dismutase:

$$2\,O_2^{-\bullet} + 2H^+ \xrightarrow{\text{superoxide dismutase}} H_2O_2 + O_2$$

Despite the ability of $O_2^{-\bullet}$ and H_2O_2 to react with a number of biologic substrates in cell-free systems, neutrophils appear to be unable to utilize either of these metabolites alone to incite extracellular injury. Under physiologic conditions, the dismutation reaction is extremely efficient and thereby ensures that the majority of $O_2^{-\bullet}$ produced by NADPH oxidase systems is rapidly consumed in the formation of H_2O_2.[98] In addition, most of the H_2O_2 produced by PMNs is catabolized by neutrophils themselves, and little H_2O_2 is released into the extracellular space.[110] Thus, despite the reactivity of both $O_2^{-\bullet}$ and H_2O_2, there is little evidence that neutrophils utilize either of these oxygen metabolites to directly attack tissues subjected to ischemia-reperfusion. As a result, considerable attention has been devoted to the possibility that much of the oxidative injury produced by activated neutrophils depends upon the participation of these metabolites in the elaboration of other, more potent, oxygen-derived radicals, such as the hydroxyl radical ($OH^\bullet$). The formation of $OH^\bullet$ in the presence of $O_2^{-\bullet}$ and H_2O_2 is described by the Haber-Weiss equation:[111]

$$O_2^{-\bullet} + H_2O_2 \longrightarrow OH^\bullet + OH^- + O_2$$

Under physiologic conditions, the Haber-Weiss reaction proceeds too slowly to be of biologic significance;[112] however, in the presence of a transition metal catalyst (M), $OH^\bullet$ may be generated more rapidly through Fenton chemistry, wherein $O_2^{-\bullet}$ first reacts with the oxidized form of a trace metal, resulting in reduction of the metal and the formation of molecular oxygen.[99] The reduced metal cation may then react with H_2O_2 to form OH^- and hydroxyl radical, while regenerating the oxidized metal catalyst. The complete reaction is depicted below:

$$O_2^{-\bullet} + M^{(n+1)+} \longrightarrow M^{n+} + O_2$$

$$M^{n+} + H_2O_2 \longrightarrow M^{(n+1)+} + OH^\bullet + OH^-$$

$$O_2^{-\bullet} + H_2O_2 \longrightarrow OH^\cdot + OH^- + O_2$$

Ferrous iron (Fe^{2+}) is generally accepted to be the most important physiologic metal catalyst for the Haber-Weiss reaction,[113-116] and there is substantial indirect evidence linking oxidative reperfusion injury in skeletal muscle and other tissues to the iron-catalyzed reaction between $O_2^{-\bullet}$ and H_2O_2. The majority of this evidence comes in the form of studies demonstrating that the administration of hydroxyl radical scavengers (such as DMSO and mannitol)[117] or ferric iron chelators (such as deferoxamine mesylate)[118-121] protects against the cytotoxic effects of reperfusion.

Several investigations have further suggested that neutrophil-mediated endothelial cell injury is mediated by H_2O_2-dependent pathways and may involve the formation of $OH^{\cdot}$ via the iron-catalyzed Haber-Weiss reaction.[122-124]

As noted above, the generation of significant quantities of hydroxyl radical by activated neutrophils would require the participation of an endogenous Haber-Weiss catalyst. Early reports by Ambruso and Johnston[125] indicated that lactoferrin, an iron-binding protein present in the specific granules of neutrophils, was capable of promoting $OH^{\bullet}$ production by human neutrophils; however, several subsequent investigations have failed to confirm that lactoferrin is a significant source of catalytically active iron under physiologic conditions.[126-128] Neutrophils may not only lack an endogenous capacity to produce $OH^{\bullet}$, but their release of lactoferrin and myeloperoxidase may actually inhibit $OH^{\bullet}$ formation.[129,130] Consequently, the generation of $OH^{\bullet}$ by neutrophils in vivo would require both the involvement of an exogenous Haber-Weiss catalyst and the ability to overwhelm the inhibitory effects of neutrophil lactoferrin and myeloperoxidase. Gannon et al[131] have demonstrated that the iron is essential to neutrophil-dependent oxygen radical-mediated endothelial cell injury and that the iron necessary for neutrophil oxidant production is derived from the endothelial cells themselves. In their experiments, neutrophils stimulated in the presence of deferoxamine were noted to undergo a normal respiratory burst and to exhibit unimpaired cytotoxicity for endothelial cells. Pretreatment of the target endothelial cells with deferoxamine, however, resulted in a high degree of protection from neutrophil-mediated cytotoxicity. The investigators propose that stimulation of neutrophils by the appropriate proinflammatory mediators results in the activation of membrane-associated NADPH oxidase and the subsequent generation of $O_2^{-\bullet}$ and H_2O_2. At the same time, iron is liberated from intracellular ferritin stores, perhaps as a direct result of the reducing action of $O_2^{-\bullet}$ that has entered the cell through small anionic channels. Reduced iron may then catalyze the formation of $OH^{\bullet}$ from H_2O_2. In this schema, target cells provide a catalyst that is critical to the pathogenesis of their own destruction.

There is considerable experimental evidence that ischemia-reperfusion of skeletal muscle activates the neutrophil NADPH oxidase system and that the subsequent production of oxygen-derived free radicals plays an important role in the genesis of postischemic tissue injury.[7,132] Neutrophils collected from the venous effluent of reperfused animal hindlimbs display an enhanced respiratory burst compared to preischemic levels and are primed to generate increased quantities of reactive oxygen metabolites following stimulation.[7] Cambria et al[132] have documented the activation of neutrophil NADPH oxidase by measuring the liberation of $O_2^{-\bullet}$ from polymorphonuclear leukocytes isolated from the gracilis muscle venous effluent and from central venous blood following 6 hours of bilateral canine gracilis ischemia and 1 hour of reperfusion. Neutrophils isolated from baseline samples drawn prior to ischemia produced minimal $O_2^{-\bullet}$ without stimulation, 0.33 ± 0.15 nmole of $O_2^{-\bullet}/2 \times 10^6$ PMNs, whereas stimulation of these cells with opsonized zymosan increased production of $O_2^{-\bullet}$ to 10.3 ± 1.4 nmole. Postischemic samples taken from both the gracilis vein and the central venous blood showed a significant increase in $O_2^{-\bullet}$ production compared to preischemic samples. Unstimulated neutrophils isolated from the gracilis vein at the end of the reperfusion period yielded 0.84 ± 0.16 nmole of $O_2^{-\bullet}$, and those isolated from central venous blood generated 0.96 ± 0.08 nmole ($P < 0.01$). Stimulation of the cells with opsonized zymosan increased $O_2^{-\bullet}$ production to 15.5 ± 0.9 nmole ($P < 0.01$) and 15.2 ± 1.1 nmole ($P < 0.05$) respectively. The elaboration of $O_2^{-\bullet}$ from both stimulated and unstimulated neutrophils isolated from the central venous blood of animals subjected to sham ischemia was not significantly different from that measured at baseline. These results indicate that neutrophil NADPH oxidase systems are activated during skeletal muscle ischemia-reperfusion and that circulating neutrophils are "primed" in vivo to respond to external stimuli with an enhanced production of toxic oxygen metabolites. Although the exact nature of the priming stimulus is uncertain, the presence of activated neutrophils in both the local venous effluent and in the central venous blood suggests that leukocyte activation following ischemia is a result of the action of soluble proinflammatory mediators, such as products of the complement cascade or of arachidonic acid metabolism. The systemic activation of circulating neutrophils may also explain the leukosequestration that may be observed in remote tissues, such as the pulmonary microvasculature, following hindlimb skeletal muscle ischemia.

It has been recognized that the administration of proinflammatory agents that stimulate the neutrophil respiratory burst can cause an adherence of polymorphonuclear leukocytes to the microvascular endothelium and a significant increase in skeletal muscle vascular resistance. Activation of circulating neutrophils by the intravascular administration of zymogen-activated plasma (ZAP), an agent which stimulates neutrophils to generate oxygen-derived free radicals through complement activation, is associated with a dramatic increase in microvascular permeability, as assessed by measurement of the solvent drag coefficient for plasma proteins, and tissue neutrophil infiltration, as quantitated by myeloperoxidase activity, in canine gracilis muscle preparations.[133] The changes in vascular permeability observed after ZAP infusion are effectively abolished in animals rendered neutropenic by antineutrophil serum as well in those pretreated with the anti-CD18 monoclonal antibody IB_4, the iron chelator deferoxamine, or catalase, although only neutropenia and inhibition of neutrophil adherence prevent ZAP-induced leukosequestration. In contrast, inhibition or inactivation of the superoxide-generating enzyme xanthine oxidase provides no significant protection against ZAP-induced microvascular dysfunction.[133] These findings indicate that complement-activated neutrophils produce microvascular dysfunction in skeletal muscle through the generation of cytotoxic oxygen metabolites. Moreover, leukocyte-dependent microvascular dysfunction appears to result from the elaboration of oxygen-derived free radicals through pathways that do not involve xanthine oxidase. The etiologic role of neutrophil-derived oxidants in skeletal muscle injury induced by ischemia-reperfusion is similarly suggested by studies in which the administration of oxygen free radical scavengers affords protection equivalent to that provided by leukocyte depletion.[4] Following 90 minutes of high-grade partial hindlimb ischemia, for example, the salutary effect of pretreatment with superoxide dismutase and catalase on the repolarization of resting E_m is similar to that achieved by neutrophil depletion alone, implicating polymorphonuclear leukocytes as the source of free radicals responsible for cell membrane injury following skeletal muscle ischemia[4] (Fig. 3.5). In the same model, the administration of the xanthine oxidase inhibitor allopurinol fails to provide significant protection against reperfusion-associated membrane dysfunction, indicating that neutrophil-dependent oxidative skeletal muscle injury induced by ischemia-reperfusion is similar to that produced by ZAP-activated neutrophils in that it does not appear to involve the activity of xanthine oxidase.

The observation that treatment with oxygen free radical scavengers is as effective as leukocyte depletion or prevention of neutrophil adherence in attenuating skeletal muscle ischemia-reperfusion injury indicates that the elaboration of reactive oxygen metabolites is an essential component of neutrophil-mediated injury in postischemic skeletal muscle. It should be emphasized, however, the ability of neutrophil-derived oxidants to directly injure postischemic tissues has yet to be definitively established, and it has been suggested that reactive oxygen metabolites may play only an indirect role in the genesis of skeletal muscle reperfusion injury. Exposure of arachidonic acid to superoxide anion in vitro results in the production of a lipid with potent chemotactic activity for neutrophils.[134] Human plasma exposed to superoxide anion becomes similarly chemotactic,[135] suggesting that the ability of superoxide dismutase to inhibit the migration of neutrophils to sites of inflammatory challenge results from its ability to prevent the formation of superoxide-generated chemoattractants. Recent studies have demonstrated that superoxide anion may also play a critical role in mediating neutrophil adhesion to the postischemic microvascular endothelium.[136,137] Thus, it is possible that the deleterious effects of oxidants produced following ischemia-reperfusion may arise not from their direct cytotoxicity, but rather from their ability to promote the accumulation of activated leukocytes in postischemic tissues.

Myeloperoxidase and the Production of HOCl

Myeloperoxidase is an enzyme that is present in abundance within the cytoplasmic granules of polymorphonuclear leukocytes. By reacting with H_2O_2, myeleloperoxidase oxidizes the halides Cl^-, Br^-, and I^- to their corresponding hypohalous acids (HOX):[98,104]

$$H_2O_2 + X^- + H^+ \xrightarrow{\text{myeloperoxidase}} HOX + H_2O$$

Because of the abundance of chloride in the extracellular fluid, the predominant product of neutrophil myeloperoxidase is hypochlorous acid (HOCl),[98] an extremely potent oxidant which can attack a number of biologically relevant targets including amines, amino acids, thiols, thioesters, nucleotides, hemoproteins, and polyenoic acids.[138] Neutrophil activation results in the elaboration of large quantities of HOCl, and in vitro quantitative analyses have demonstrated that the HOCl generated by stimulated neutrophils accounts for the catalysis of nearly all of the H_2O_2 produced by the cells.[138]

In rats, instillation of H_2O_2 and myeloperoxidase into the lungs or renal arteries

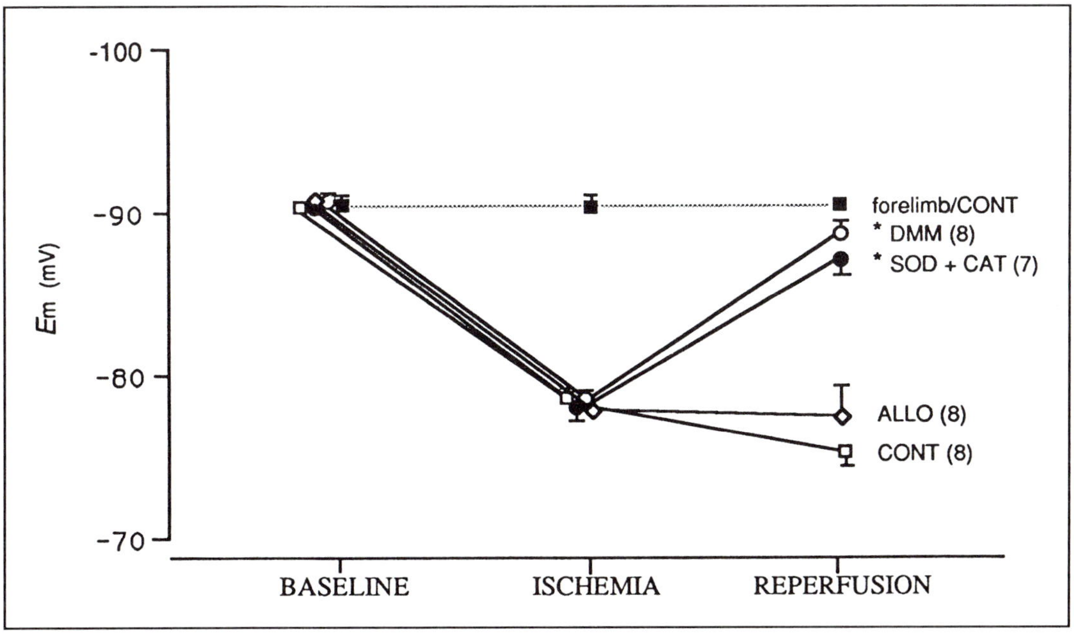

*Fig. 3.5. Hindlimb muscle transmembrane potential (Em) in control and treated groups (solid lines) and forelimb muscle Em in control group (dotted line) at baseline, after 90 minutes of ischemia, and after 60 minutes of reperfusion. Values shown are means ± SEM. * P < 0.05 vs. hindlimb control during reperfusion. Hindlimb Em in all groups was significantly depolarized (P < 0.05 vs. baseline) after 90 minutes of ischemia. Hindlimb Em during reperfusion was significantly different from that measured during ischemia in animals rendered neutropenic with dimethylmyleran (DMM) and in those treated with the free radical scavengers superoxide dismutase and catalase (SOD + CAT), whereas hindlimb Em during reperfusion was not significantly different from that measured during ischemia in the untreated control (CONT) and allopurinol (ALLO) groups.*

results in severe pulmonary or glomerular injury, respectively.[139,140] The protective effect of the H_2O_2-scavenging enzyme catalase, which removes an essential substrate for neutrophil-derived myeloperoxidase, suggests a similar role for HOCl in the genesis of postischemic skeletal muscle dysfunction as well.[117] However, several studies have indicated that chlorinated oxidants have only a limited ability to mediate toxicity in mammalian systems,[98] and, despite the high oxidative potential of myeloperoxidase products, their ability to cause direct cellular injury in tissues subjected to ischemia-reperfusion remains unproven.

Proteolytic Enzymes

In contrast to the action of unstable cytotoxic oxygen radicals, the destructive capacity of proteolytic enzymes can be exceptionally long-lived and is invariably directed against relatively specific substrates. Yet despite the destructive potential of the granular enzymes of neutrophils, their role in mediating ischemia-reperfusion injury in skeletal muscle has been less well studied than that of reactive oxygen species. Of the more than 20 distinct proteolytic enzymes contained within neutrophil granules, three have received the most attention with respect to their ability to induce tissue injury. These are the serine protease, elastase, and two metalloproteinases, collagenase and gelatinase. The toxicity of these enzymes arises from their ability to attack and degrade essential components of the extracellular matrix including elastin, collagen, glycoproteins and proteoglycans. No longer regarded simply as an inert scaffold for living cells, the extracellular matrix has been shown to serve an essential function in the maintenance, growth, and repair of tissues.[141,142] Degradation of the extracellular matrix by neutrophil proteases may therefore result in significant alterations of cell structure and function.

Elastase is present in active form within the azurophilic granules of neutrophils and is capable of hydrolyzing a variety of biological substrates including types III and IV collagen, fibronectin, proteoglycans, immunoglobulins, complement proteins, and even intact cells.[143,144] Its primary physiologic function is believed to be limited to the intracellular compartment, where it participates in the destruction of microorganisms contained within phagocytic vacuoles.[143] The ability of elastase to mediate extracellular damage is limited by the presence of several potent antiproteinases in the plasma and interstitial fluids, including α_1-antiproteinase inhibitor (formerly termed α_1-antitrypsin), α_2-macroglobulin, and secretory leukoproteinase inhibitor.[145] The most important defense against extracellular elastase activity is α_1-antiproteinase inhibitor, which efficiently and irreversibly binds to elastase, forming an inactive enzyme-inhibitor complex.[145]

It would seem unlikely that neutrophil elastase would have the opportunity to injure host tissues in the presence of such an efficient and redundant antiproteinase defense system. However, there is substantial evidence that the extracellular antiproteinases may be rendered inactive in the presence of inflammation, allowing elastase to escape destruction and to injure host tissues.[98,145] The mechanism underlying the inactivation of the antiproteinase defense system was defined in a series of experiments conducted in the 1980s. Structural evaluation of α_1-antiproteinase inhibitor revealed that the molecule contains a critical methionine residue in its active site (position 358) that is susceptible to oxidation by chemical reagents or activated neutrophils. Oxidation of Met-358 results in a marked decrease in the affinity of α_1-antiproteinase inhibitor for neutrophil elastase and increases the in vivo half-life of elastase 2000-fold.[143,145] The cooperative use of oxygen metabolites and lysosomal enzymes to attack and degrade the extracellular matrix was demonstrated by Weiss et al[146] in an in vitro model involving the activation of neutrophils upon a matrix of connective tissue macromolecules in the presence of purified α_1-antiproteinase inhibitor. Under the conditions of the study, oxidation of α_1-antiproteinase inhibitor by neutrophil-generated HOCl resulted in inactivation of the inhibitor and solubilization of the connective tissue matrix by neutrophil elastase. The other antiproteinases, α_2-macroglobulin and secretory leukoproteinase inhibitor, have also been shown to be susceptible to oxidative inactivation by neutrophils.[98,147] Thus, by releasing oxidants into the extracellular space, the neutrophil may create a local microenvironment in which antiproteinases are inactivated and elastase can act freely to degrade susceptible host tissues.

Elastase has been implicated in the pathophysiology of several diseases, including emphysema, adult respiratory distress syndrome, and glomerulonephritis. In addition, elevated plasma elastase activity has been detected in patients with myocardial ischemia[148,149] and in those undergoing thrombolytic treatment for acute myocardial infarction.[150] Release of elastase from neutrophils appears to be essential for the diapedesis of activated leukocytes into the interstitium of tissues exposed to ischemia-reperfusion.[151] In addition, in vitro experiments suggest that elastase may play a significant etiologic role in anoxia-reoxygenation-induced, neutrophil-mediated endothelial dysfunction.[152-154] Inauen et al[152] reported that exposure of cultured bovine microvascular endothelial cells to anoxia-reperfusion resulted in endothelial cell injury, as indicated by cell lysis and detachment from the underlying matrix, and that the addition of neutrophils upon reoxygenation significantly exacerbated endothelial cell injury. Moreover, exposure of the endothelial cells to anoxia induced the release of elastase from neutrophils added upon reoxygenation. In the presence of specific inhibitors of neutrophilic elastase, leukocyte-mediated cell detachment was significantly attenuated, suggesting that elastase may be an important mediator of nonlytic neutrophil-induced endothelial cell injury following anoxia-reoxygenation.

The metalloproteinases, collagenase and gelatinase, share with elastase the ability to degrade essential components of the extracellular matrix. Neutrophil collagenase has been shown cleave each of the interstitial collagens (types I, II, and III), whereas gelatinase may attack types V and XI, as well as type IV, an essential component of the basement membrane.[155] Unlike elastin, the metalloproteinases are synthesized and released as latent, inactive proenzymes.[155] For many years, the physiologic mechanism by which neutrophils activated their metalloproteinases remained unknown. Whereas exposure to the proteolytic action of trypsin was found to result in the activation of neutrophil collagenase in vitro, treatment with the neutrophil-derived serine proteases elastase and cathepsin G was noted to destroy the proenzyme.[155] The ability of neutrophil-derived oxidants to activate the latent form of collagenase was first suggested by studies of neutrophils isolated from patients with chronic granulomatous disease (CGD), a disorder characterized by a defect in neutrophil NADPH oxidase which renders the cells unable to generate significant quantities of $O_2^{-\bullet}$, H_2O_2 , and HOCl. Following stimulation, both normal and CGD neutrophils discharged collagenase into the extracellular space; however, unlike normal neutrophils, those isolated from patients with CGD were unable to activate the proenzyme, indicating that neutrophil oxidant production was a prerequisite to collagenase activation.[156] Further investigation revealed that the ability of normal neutrophils to activate latent collagenase could be inhibited if the cells were specifically prevented from generating HOCl and that the addition of exogenous HOCl to cell-free preparations of latent collagenase was sufficient to cause activation of the enzyme.[156] A similar role for HOCl has been demonstrated with regard to the activation of latent neutrophil gelatinase.[157] Thus, like that of elastase, the activities of neutrophil collagenase and gelatinase appear to be closely linked to the generation of HOCl by the neutrophil myeloperoxidase system. Such a cooperative interaction between neutrophil-derived oxidants and proteolytic enzymes may serve to enhance the ability of neutrophils to degrade host tissues (Fig. 3.6).

THE NO-REFLOW PHENOMENON

More than half a century ago, Harman[158] recognized that a significant portion of the skeletal muscle microvasculature failed to perfuse following the removal of tourniquets from animal hindlimbs and subsequently proposed that capillary no-reflow might be

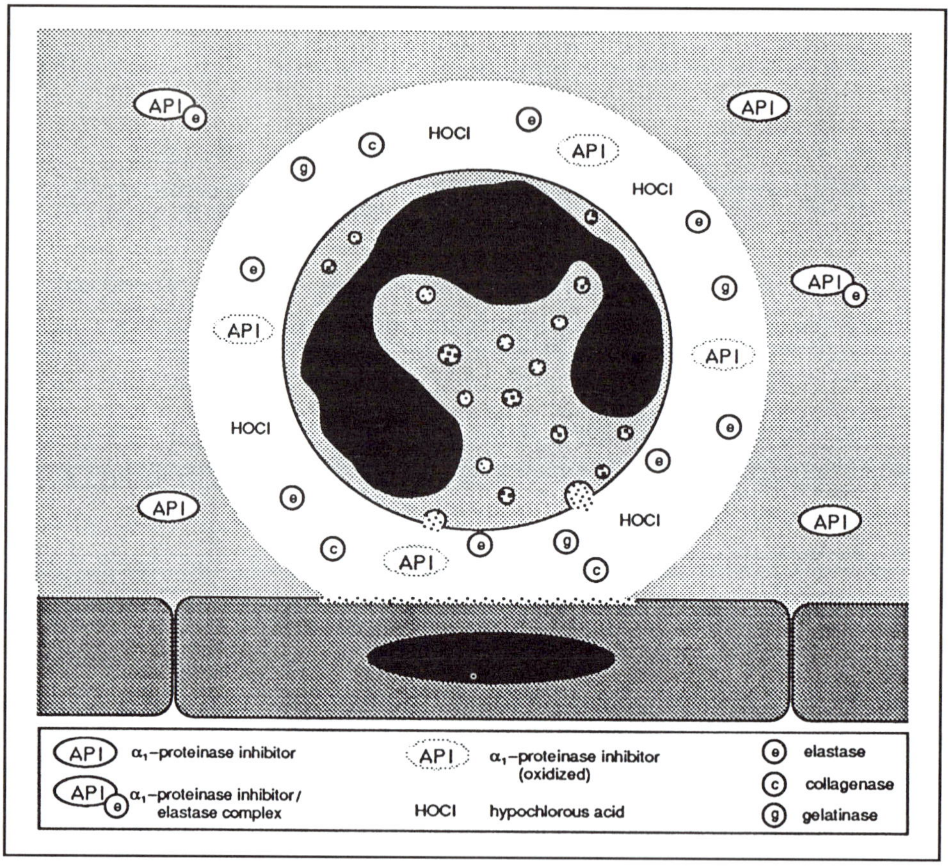

Fig. 3.6. Schematic representation of the cooperative interaction between neutrophil-generated oxidants and proteolytic enzymes. Release of myeloperoxidase-derived hypochlorous acid (HOCl) by activated neutrophils creates a zone of oxidized, inactivated α_1-proteinase inhibitor (API) in which elastase may act in an unregulated fashion, injuring host tissues in the local microenvironment. The latent metalloenzymes collagenase and gelatinase are released from the neutrophil and are oxidatively activated in the presence of HOCl, resulting in the degradation of extracellular matrix components. Outside of this zone of neutrophil-mediated injury, unoxidized α_1-proteinase inhibitor may bind to and inactivate released neutrophil elastase, thus limiting the domain of its destructive potential.

the principal cause of skeletal muscle necrosis following ischemia. Strock and Majno[159,160] later confirmed the occurrence of perfusion defects in the microcircularion of postischemic skeletal muscle and suggested that leukocyte trapping might be a potential mechanism for the impairment of capillary reflow that was observed after skeletal muscle ischemia. It is now widely believed that capillary plugging by neutrophils plays a significant role in the pathogenesis of the no-reflow phenomenon, and along with the elaboration of reactive oxygen metabolites and the release of proteolytic lysosomal enzymes may be an important cause of neutrophil-mediated skeletal muscle injury following ischemia-reperfusion.

To some degree, the ability of neutrophils to interfere with microcirculatory flow may be attributed to their physical characteristics. In contrast to erythrocytes, neutrophils are relatively large, stiff viscoelastic cells.[161] In order to pass through the microvasculature, they must undergo considerable deformation.[161] Because of their biophysical properties and their natural tendency to adhere to the microvascular endothelium, neutrophils entering the microcirculation can impose a substantial resistance to capillary blood flow. Moreover, the contribution of neutrophils to microvascular resistance may become exaggerated under certain pathologic conditions, such as ischemia-reperfusion, that cause cell activation, cytoplasmic stiffening, and adhesion to the microvascular endothelium. Whole organ quantitative perfusion studies have indicated that leukocytes normally account for approximately 22% of the whole blood resistance in skeletal muscle.[162,163] Activation of leukocytes increases their effect to 50-60% of the total resistance.[163] In a recent study, Harris and Skalak[164] demonstrated that under normal physiologic conditions capillary plugging by leukocytes contributed only 1.1% of the total resistance to microvascular flow in skeletal muscle. When the leukocytes were activated by superfusion with *N*-formyl-methionyl-leucyl-phenylalanine (FMLP), there was approximately a 23-fold increase in this contribution to the total resistance, suggesting that capillary plugging by activated leukocytes may significantly impair microcirculatory flow.

Direct evidence of leukocyte plugging as an etiologic factor in capillary flow maldistribution during skeletal muscle ischemia has been reported by Bagge et al.[12] In their experiments, intravital microscopy of the feline tenuissimus muscle during hemorrhagic shock revealed a reduction in the number of perfused capillaries by approximately 50%. Close examination of the nonperfused capillaries showed that most remained patent and almost invariably contained one or more leukocytes. Spontaneous dislodgement of the trapped leukocytes resulted in the prompt restoration of capillary flow, supporting the notion that leukocyte plugging may play an important role in the genesis of the no-reflow phenomenon in skeletal muscle during hemorrhagic shock. Lehr et al[29] later observed that reperfusion of ischemic skeletal muscle in the hamster dorsal skinfold elicited the firm adhesion of leukocytes to the endothelium of postcapillary venules. Throughout the reperfusion period, the functional capillary density, i.e., the density of erythrocyte-perfused capillaries, of the muscle was significantly reduced, indicating that plugging of the capillary lumen by adherent leukocytes had resulted in the impairment of microcirculatory flow.

Further support for the role of neutrophils in the development of microvascular perfusion impairment following ischemia is provided by the observation that neutrophil depletion significantly attenuates the extent of capillary no-reflow in postischemic brain[165] and myocardium.[166] Neutropenia has similarly been shown to similarly diminish capillary obstruction in several tissues following the global ischemia of hemorrhagic shock.[11] Studies demonstrating that leukocyte depletion and prevention of leukocyte adherence dramatically reduce the increase in skeletal muscle vascular resistance following ischemia-reperfusion provide additional evidence that activated neutrophils may play a role in postischemic microvascular obstruction.[3,18] Neutrophil depletion with specific polyclonal antineutrophil serum has recently been shown to significantly improve postischemic microvascular patency in canine gracilis muscle subjected to 4 hours of ischemia and 30 minutes of reperfusion.[13] In the same model, preservation of capillary flow is also provided by pretreatment with the anti-CD18 monoclonal antibody IB_4, indicating that the development of no-reflow in postischemic skeletal muscle is dependent upon the integrin-mediated interaction between neutrophils and the microvascular endothelium.[13] Since previous investigations demonstrated that pretreatment with antineutrophil serum or monoclonal antibody IB_4 abolished the increase in microvascular permeability and the development of edema after canine gracilis ischemia-reperfusion,[18] it is possible that the salutory effects of these agents on capillary patency in this model may result from a decrease in the extent of extravascular compression arising from the accumulation of interstitial fluid or the swelling of postischemic myocytes rather than from intraluminal obstruction. Thus, although integrin-mediated interactions may strengthen the attachment of neutrophils to the microvascular endothelium, thereby increasing the likelihood that leukocytes will become trapped within the capillary lumen, the exact role of CD18-mediated leukocyte adherence in the development of intraluminal capillary obstruction following skeletal muscle ischemia-reperfusion has yet to established.

CONCLUSIONS

While the beneficial effects of the inflammatory response on host defense and repair are indisputable, it is now widely recognized that leukocyte activation during pathological states such as ischemia-reperfusion can result in the release of toxic products into the extracellular space with the consequent exacerbation of tissue injury. Investigations into the pathophysiology of skeletal muscle ischemia-reperfusion have revealed a process characterized by granulocyte adhesion to the microvascular endothelium and the subsequent emigration of activated leukocytes from the vascular lumen toward chemotactic stimuli within the extravascular tissues. Extravasated neutrophils may thereafter release a variety of cytotoxic substances into the extracellular space, including reactive oxygen metabolites, lysosomal enzymes, and products of arachidonic acid metabolism, such as prostaglandins and leukotrienes. These substances are potent mediators of the microvascular and cellular effects of inflammation and may significantly amplify the deleterious effects of the initial ischemic insult. In addition, neutrophils may become trapped within the microvasculature and contribute to the development of the no-reflow phenomenon, thereby exacerbating the extent of ischemic injury despite the apparent restoration of blood flow.

Although it was initially assumed that the microvasculature served simply as a conduit for the movement of leukocytes into injured tissues, it is now understood that the vascular endothelium plays an active role in modulating neutrophil activity. Recent investigations indicate that the upregulation of specific adhesion molecules on the endothelial surface is a critical step in the infiltration of neutrophils into postischemic tissues. Moreover, experimental evidence suggests that the elaboration of proinflammatory mediators by stimulated endothelial cells may serve an important function in the process of neutrophil activation and migration following ischemia-reperfusion. The exact source and identity of the chemotactic signal or signals responsible for leukosequestration in postischemic skeletal muscle remain uncertain, as does the nature of the biologically relevant mediators of adhesion molecule expression on the surface of activated leukocytes and endothelial cells. In addition, the precise role of neutrophil-derived free radicals and proteolytic enzymes in the generation of postischemic microvascular dysfunction and skeletal muscle injury is still incompletely understood. Thus, despite the remarkable advances that have been made in the past decade, the topic of skeletal muscle ischemia-reperfusion injury remains a fertile area for further investigation.

REFERENCES

1. Dahlbäck L, Rais O. Morphologic changes in striated muscle following ischemia: immediate postischemic phase. Acta Chir. Scand. 1966; 131:430-440.
2. Walden DL, McCutchan HJ, Enquist EG et al. Neutrophils accumulate and contribute to skeletal muscle dysfunction after ischemia-reperfusion. Am J. Physiol. 1990; 259:H1809-H1812.
3. Korthuis RJ, Grisham MB, Granger DN. Leukocyte depletion attenuates vascular injury in postischemic skeletal muscle. Am. J. Physiol. 1988; 254(Heart Circ. Physiol. 23):H823-H827.
4. Yokota J, Minei JP, Fantini GA et al. Role of leukocytes in reperfusion injury of skeletal muscle after partial ischemia. Am. J. Physiol. 1989; 257(Heart Circ. Physiol. 26):H1068-H1075.
5. Belkin M, LaMorte ML, Wright JG et al. The role of leukocytes in the pathophysiology of skeletal muscle ischemic injury. J. Vasc. Surg. 1989; 10:14-17.
6. Freischlag JA, Hanna D. Neutrophil (PMN) phagocytosis and chemotaxis after 2 hr of ischemia. J. Surg. Res. 1991; 50:648-652.
7. Goldman G, Welbourn R, Klausner JM et al. Ischemia activates neutrophils but inhibits their local and remote diapedesis. Ann. Surg. 1990; 211:196-201.
8. Goldman G, Welbourn R, Paterson IS et al. Ischemia-induced neutrophil activation

and diapedesis is lipoxygenase dependent. Surgery 1990; 107:428-433.
9. Cambria RA, Anderson RJ, Dikdan G et al. The influence of arachidonic acid metabolites on leukocyte activation and skeletal muscle injury after ischemia-reperfusion. J. Vasc. Surg. 1991; 14:549-556.
10. Klausner JM, Paterson IS, Valeri CR et al. Limb ischemia-induced increase in permeability is mediated by leukocytes and leukotrienes. Ann. Surg. 1988; 208: 755-760.
11. Barroso-Aranda J, Schmid-Schönbein GW, Zwiefach BW et al. Granulocytes and noreflow phenomenon in irreversible hemorrhagic shock. Circ. Res. 1988; 63:437-447.
12. Bagge U, Amundson B, Lauritzen C. White blood cell deformability and plugging of skeletal muscle capillaries in hemorrhagic shock. Acta Physiol. Scand. 1980; 180: 159-163.
13. Jerome SN, Smith CW, Korthuis RJ. CD18-dependent adherence reactions play an important role in the development of the no-reflow phenomenon. Am. J. Physiol. 1993; 264(Heart Circ. Physiol. 33): H479-H483.
14. Romson JL, Hook BG, Kunkel SL et al. Reduction of the extent of acute myocardial injury by neutrophil depletion in the dog. Circulation 1983; 67:1016-1023.
15. Mullane KM, Read N, Salmon JA et al. Role of leukocytes in acute myocardial infarction in anesthetized dogs. J. Pharmacol Exp. Ther. 1984; 228:510-522.
16. Mallory G, White P, Salcedo-Salger J. The speed of healing of myocardial infarction: a study of the pathologic anatomy in seventy-two cases. Am. Heart J. 1939; 18:647-671.
17. Smith JK, Grisham MB, Granger DN et al. Free radical defense mechanisms and neutrophil infiltration in postischemic skeletal muscle. Am. J. Physiol. 1989; 256: H789-H793.
18. Carden DL, Smith JK, Korthuis RJ. Neutrophil-mediated microvascular dysfunction in postischemic canine skeletal muscle: role of granulocyte adherence. Circ. Res. 1990; 66:1436-1444.
19. Roberts JP, Perry MO, Hariri RJ et al. Incomplete recovery of muscle cell function following partial but not complete ischemia. Circ Shock 1985; 17:253-258.
20. Perry MO, Shires GT III, Albert. SA. Cellular changes with graded limb ischemia and reperfusion. J Vasc Surg 1984; 1: 536-40.
21. Lorant DE, Topham MK, Whatley RE et al. Inflammatory roles of P-selectin. J. Clin. Invest. 1993; 92:559-570.
22. Shasby DM, Shasby SS, Peach MJ. Granulocytes and phorbol myristate acetate increase permeability to albumin of cultured endothelial monolayers and isolated perfused lungs: role of oxygen radicals and granulocyte adherence. Am. Rev. Resp. Dis. 1983; 127:72-79.
23. Vedder NB, Winn RK, Rice CL et al. Inhibition of leukocyte adherence by anti-CD18 monoclonal antibody attenuates reperfusion injury in the rabbit ear. Proc. Natl. Acad. Sci. USA 1990; 87:2643-2646.
24. Rosales C, Brown EJ. Neutrophil receptors and modulation of the immune response, in Abramson JS, Wheeler JG, (eds): The Neutrophil. New York, Oxford University Press, Inc., 1993, pp 23-62.
25. Tözeren A, Ley K. How do selectins mediate leukocyte rolling in venules? Biophys. J. 1992; 63:700-709.
26. Firrell JC, Lipowsky HH. Leukocyte margination and deformation in mesenteric venules of rat. Am. J. Physiol. 1989; 256 (Heart Circ. Physiol. 25): H1667-H1674.
27. Erlansson M, Bergqvist D, Persson NH et al. Modification of postischemic increases of leukocyte adhesion and vascular permeability in the hamster by iloprost. Prostaglandins 1991; 41:157-168.
28. Lehr HA, Guhlmann A, Nolte D et al. Leukotrienes as mediators in ischemia-reperfusion injury in a microcirculation model in the hamster. J. Clin. Invest. 1991; 87:2036-2041.
29. Lehr H-A, Hübner C, Nolte D et al. Dietary fish oil blocks the microcirculatory manifestations of ischemia-reperfusion injury in striated muscle in hamsters. Proc. Natl. Acad. Sci. USA 1991; 88:6726-6730.
30. Goldberg M, Serafin D, Klitzman B. Quantification of neutrophil adhesion to skeletal muscle venules following ischemia-reperfusion. J. Reconstr. Microsurg. 1990; 6:267-270.

31. Ley K. Leukocyte adhesion to vascular endothelium. J. Reconstr. Microsurg. 1992; 8:495-503.
32. Springer TA. Adhesion receptors of the immune system. Nature 1990; 346: 425-434.
33. Bevilacqua M, Butcher E, Furie B et al. Selectins: a family of adhesion receptors. Cell 1991; 67:233.
34. Ley K, Gaehtgens P, Fennie C et al. Lectin-like cell adhesion molecule 1 mediates leukocyte rolling in mesenteric venules in vivo. Blood 1991; 77:2553-2555.
35. von Adrian UH, Chambers JD, McEvoy LM et al. Two step model of leukocyte-endothelial cell interaction in inflammation: distinct roles for LECAM-1 and the leukocyte β_2 integrins in vivo. Proc. Natl. Acad. Sci. USA 1991; 88:7538-7542.
36. Abbassi O, Lane CL, Krater S et al. Canine neutrophil margination mediated by lectin adhesion molecule-1 in vitro. J. Immunol. 1991; 147:2107-2115.
37. Jutila MA, Rott L, Berg EL et al. Function and regulation of the neutrophil MEL-14 antigen in vivo: comparison with LFA-1 and Mac-1. J. Immunol. 1989; 143:3318-3324.
38. Watson SR, Fennie C, Lasky LA. Neutrophil influx into an inflammatory site inhibited by soluble homing receptor-IgG chimaera. Nature 1991; 349:164-166.
39. Lawrence MB, Springer TA. Leukocytes roll on a selectin at physiologic flow rates: Distinction from and prerequisite for adhesion through integrins. Cell 1991; 65:859-873.
40. Doré M, Korthuis RJ, Granger DN et al. P-selectin mediates spontaneous leukocyte rolling in vivo. Blood 1993; 82:1308-1316.
41. Hsu-Lin S, Berman CL, Furie BC et al. A platelet membrane protein expressed during platelet activation and secretion. Studies using a monoclonal antibody specific for thrombin-activated platelets. J. Biol. Chem. 1984; 259:9121-9126.
42. Bonfanti R, Furie BC, Furie B et al. PADGEM (GMP-140) is a component of Weibel-Palade bodies of human endothelial cells. Blood 1989; 73:1109-1112.
43. McEver RP, Beckstead JH, Moore KL et al. GMP-140, a platelet alpha-granule membrane protein, is also synthesized by vascular endothelial cells and is localized in Weibel-Palade bodies. J. Clin. Invest. 1989; 84:92-99.
44. Patel KD, Zimmerman GA, Prescott SM et al. Oxygen radicals induce human endothelial cells to express GMP-140 and bind neutrophils. J. Cell Biol. 1991; 112: 749-759.
45. Hattori R, Hamilton KK, McEver RP et al. Complement proteins C5b-9 induce secretion of high molecular weight multimers of endothelial von Willebrand factor and translocation of granule membrane protein GMP-140 to the cell surface. J. Biol. Chem. 1989; 264:9053-9060.
46. Weller A, Isenmann S, Vestweber D. Cloning of the mouse endothelial selectins. Expression of both E- and P-selectin is inducible by tumor necrosis factor. J. Biol. Chem. 1992; 267:15176-15183.
47. Gamble JR, Skinner MP, Berndt MC et al. Prevention of activated neutrophil adhesion to endothelium by soluble adhesion protein GMP-140. Science 1990; 249:414-417.
48. Geng J-G, Bevilacqua MP, Moore KL et al. Rapid neutrophil adhesion to activated endothelium mediated by GMP-140. Nature 1990; 343:757-760.
49. Bevilacqua MP, Pober JS, Mendrick DL et al. Identification of an inducible endothelial-leukocyte adhesion molecule. Proc. Natl. Acad. Sci. USA 1987; 84:9238-9242.
50. Shah SV, Baricos WH, Basci A. Degradation of human glomerular basement by stimulated neutrophils: activation of a metalloproteinase(s) by reactive oxygen metabolites. J. Clin. Invest. 1987; 79:25-31.
51. Walz G, Aruffo A, Kolanus W et al. Recognition by ELAM-1 of the sialyl-Le^x determinant on myeloid and tumor cells. Science 1990; 250:1132-1135.
52. Polley MJ, Phillips ML, Wayner E et al. CD62 and endothelial leukocyte adhesion molecule 1 (ELAM-1) recognize the same carbohydrate ligand, sialyl Lewis X. Proc. Natl. Acad. Sci. USA 1991; 88:6224-6228.
53. Phillips ML, Nudelman E, Gaeta FC et al. ELAM-1 mediates cell adhesion by recognition of a carbohydrate ligand, sialyl-Le^x. Science 1990; 250:1130-1132.
54. Bevilacqua PM, Stengelin G, Gimbrone MA Jr. et al. Endothelial leukocyte adhesion

molecule 1: an inducible receptor for neutrophils related to complement regulatory proteins and lectins. Science 1989; 243: 1160-1165.

55. Hynes RO. Integrins: a family of cell surface receptors. Cell 1987; 48:549-554.
56. Ruoslahti E. Integrins. J. Clin. Invest. 1991; 87:1-5.
57. Anderson DC, Springer TA. Leukocyte adhesion deficiency: an inherited defect in the MAC-1, LFA-1, and gp150,95 glycoproteins. Annu. Rev. Med. 1987; 38:175-194.
58. Arnaout MA. Leukocyte adhesion molecules deficiency: its structural basis, pathophysiology and implications for modulating the inflammatory response. Immunol. Rev. 1990; 114:145-180.
59. Diamond MS, Staunton DE, de Fougerolles AR et al. ICAM-1 (CD54): a counter-receptor for Mac-1 (CD11b/CD18). J. Cell. Biol. 1990; 111:3129-3139.
60. Diamond MS, Staunton DE, Marlin SD et al. Binding of the integrin Mac-1 (CD11b/CD18) to the third immunoglobulin-like domain of ICAM-1 (CD54) and its regulation by glycosylation. Cell 1991; 65: 961-971.
61. Sengelov H, Kjeldsen L, Diamond MS et al. Subcellular localization and dynamics of Mac-1 ($\alpha_m\beta_2$) in human neutrophils. J. Clin. Invest. 1993; 92:1467-1476.
62. Smith CW, Marlin SD, Rothlein R et al. Cooperative interactions of LFA-1 and Mac-1 with intercellular adhesion molecule-1 in facilitating adherence and transendothelial migration of human neutrophils in vitro. J. Clin. Invest. 1989; 83:2008-2017.
63. de la Ossa JC, Malago M, Gewertz BL. Neutrophil-endothelial cell binding in neutrophil-mediated tissue injury. J. Surg. Res. 1992; 53:103-107.
64. Hunkapillar T, Hood L. Diversity of the immunoglobulin gene superfamily. Adv. Immunol. 1989; 44:1-63.
65. Dustin ML, Springer TA. Lymphocyte function-associated antigen-1 (LFA-1) interaction with intercellular adhesion molecule-1 (ICAM-1) is one of at least three mechanisms for lymphocyte adhesion to cultured endothelial cells. J. Cell. Biol. 1988; 107:321-331.
66. Rothlein R, Dustin ML, Marlin SD et al. A human intercellular adhesion molecule (ICAM-1) distinct from LFA-1. J. Immunol. 1986; 137:1270-1274.
67. Dustin ML, Rothlein R, Bhan AK et al. Induction by IL-1 and interferon gamma: tissue distribution, biochemistry, and function of a natural adherence molecule (ICAM-1). J. Immunol. 1986; 137:245-254.
68. Staunton DE, Dustin ML, Springer TA. Functional cloning of ICAM-2, a cell adhesion ligand for LFA-1 homologous to ICAM-1. Nature 1989; 339:61-64.
69. Winn RK, Liggitt D, Vedder NB et al. Anti-P-selectin monoclonal antibody attenuates reperfusion injury to the rabbit ear. J. Clin. Invest. 1993; 92:2042-2047.
70. McEver RP. GMP-140: a receptor for neutrophils and monocytes on activated platelets and endothelium. J. Cell. Biochem. 1991; 45:156-161.
71. Vedder NB, Harlan JM. Increased surface expression of CD11b/CD18 (Mac-1) is not required for stimulated neutrophil adherence to cultured endothelium. J. Clin. Invest. 1988; 81:676-682.
72. Luscinskas FW, Cybulsky MI, Kiely J-M et al. Cytokine-activated human endothelial monolayers support enhanced neutrophil transmigration via a mechanism involving both endothelial-leukocyte adhesion molecule-1 and intercellular adhesion molecule-1. J. Immunol. 1991; 146:1617-1625.
73. Weyrich AS, Ma X-L, Albertine KH et al. In vivo neutralization of P-selectin protects feline heart and endothelium in myocardial ischemia and reperfusion injury. J. Clin. Invest. 1993; 91:2620-2629.
74. Oliver MG, Specian RD, Perry MA et al. Morphologic assessment of leukocyte-endothelial cell interactions in mesenteric venules subjected to ischemia and reperfusion. Inflammation 1991; 15:331-346.
75. Yoshida N, Granger DN, Anderson DC et al. Anoxia/reoxygenation-induced neutrophil adherence to cultured endothelial cells. Am. J. Physiol. 1992; 262 (Heart Circ. Physiol. 31):H1891-H1898.
76. Arfors K-E, Lundberg C, Lindbom L et al. A monoclonal antibody to the membrane glycoprotein complex CD18 inhibits polymorphonuclear leukocyte accumulation and plasma leakage in vivo. Blood 1987; 69:338-340.

77. Weselcouch EO, Grove RI, Demusz CD et al. Effect of in vivo inhibition of neutrophil adherence on skeletal muscle function during ischemia in ferrets. Am. J. Physiol. 1991; 261(Heart Circ. Physiol. 30): H1178-H1183.
78. Kukielka GL, Hawkins HK, Michael L et al. Regulation of intercellular adhesion molecule-1 (ICAM-1) in ischemic and reperfused canine myocardium. J. Clin. Invest. 1993; 92:1504-1516.
79. Ma X-L, Lefer DJ, Lefer AM et al. Coronary endothelial and cardiac protective effects of a monoclonal antibody to intercellular adhesion molecule-1 in myocardial ischemia and reperfusion. Circulation 1992; 86: 937-946.
80. Horgan MJ, Ge M, Gu J et al. Role of ICAM-1 in neutrophil-mediated lung vascular injury after occlusion and reperfusion. A. J. Physiol. 1991; 261(Heart Circ. Physiol. 30):H1578-H1584.
81. Palmer RMJ, Moncada S. A novel citrulline-forming enzyme implicated in the formation of nitric oxide by vascular endothelial cells. Biochem. Biophys. Res. Commun. 1989; 158:348-352.
82. Kubes P, Suzuki M, Granger DN. Nitric oxide: an endogenous modulator of leukocyte adhesion. Proc. Natl. Acad. Sci. USA 1991; 88:4651-4655.
83. Gaboury J, Woodman RC, Granger DN et al. Nitric oxide prevents leukocyte adherence: role of superoxide. Am. J. Physiol. 1993; 265(Heart Circ. Physiol. 34): H862-H867.
84. Tsao PS, Aoki N, Lefer DJ et al. Time course of endothelial dysfunction and myocardial injury during myocardial ischemia and reperfusion in the cat. Circulation 1990; 82:1402-1412.
85. Lefer AM, Tsao PS, Lefer DJ et al. Role of endothelial dysfunction in the pathogenesis of reperfusion injury after myocardial ischemia. FASEB J. 1991; 5:2029-2034.
86. Weyrich AS, Ma X-L, Lefer AM. The role of L-Arginine in ameliorating reperfusion injury after myocardial ischemia in the cat. Circulation 1992; 86:279-288.
87. Siegfried MR, Carey C, Ma X-L et al. Beneficial effects of SPM-5185, a cysteine-containing NO donor in myocardial ischemia-reperfusion. Am. J. Physiol. 1992; 263 (Heart Circ. Physiol. 32):H771-H777.
88. Siegfried MR, Erhardt J, Rider T et al. Cardioprotection and attenuation of endothelial dysfunction by organic nitric oxide donors in myocardial ischemia-reperfusion. J. Pharmacol. Exp. Ther. 1992; 260: 668-675.
89. Johnson G III, Tsao PS, Lefer AM. Cardioprotective effects of authentic nitric oxide in myocardial ischemia with reperfusion. Crit. Care Med. 1990; 19:244-252.
90. Cronstein BN, Levin RI, Belanoff J et al. Adenosine: an endogenous mediator of neutrophil-mediated injury to endothelial cells. J. Clin. Invest. 1986; 78:760-770.
91. Cronstein BN, Kramer SB, Weissman G et al. Adenosine: a physiological modulator of superoxide anion generation by human neutrophils. J. Exp. Med. 1983; 158: 1160-1173.
92. Cronstein BN, Rosenstein ED, Kramer SB et al. Adenosine: a physiologic modulator of superoxide anion generation by human neutrophils. Adenosine acts via an A_2 receptor of human neutrophils. J. Immunol. 1985; 135:1366-1371.
93. Cronstein BN, Kubersky SM, Weissman G et al. Engagement of adenosine receptors inhibits hydrogen peroxide release by activated human neutrophils. Clin. Immunol. Immunopathol. 1987; 42:76-85.
94. Cronstein BN, Levin RI, Philips M et al. Neutrophil adherence to endothelium is enhanced via adenosine A_1 receptors and inhibited via adenosine A_2 receptors. J. Immunol. 1992; 148:2201-2206.
95. Grisham MB, Hernandez LA, Granger DN. Adenosine inhibits ischemia-reperfusion-induced leukocyte adherence and extravasation. Am. J. Physiol. 1989; 257(Heart Circ. Physiol. 26):H1334-H1339.
96. Nolte D, Lehr H-A, Messmer K. Adenosine inhibits postischemic leukocyte-endothelium interaction in postcapillary venules. Am. J. Physiol. 1991; 261(Heart Circ. Physiol. 30):H651-H655.
97. Kitakaze M, Hori M, Morioka T et al. Attenuation of ecto-5'-nucleotidase activity and adenosine release in activated human polymorphonuclear leukocytes. Circ. Res. 1993; 73:524-533.

98. Weiss SJ. Tissue destruction by neutrophils. N. Engl. J. Med. 1989; 320(6):365-376.
99. Halliwell B, Gutteridge JMC. Oxygen toxicity, oxygen radicals, transition metals and disease. Biochem. J. 1984; 219:1-14.
100. Freeman BA, Crapo JD. Biology of disease: free radicals and tissue injury. Lab. Invest. 1982; 47:412-426.
101. Del Maestro RF. An approach to free radicals in medicine and biology. Acta Physiol. Scand. [Suppl.] 1980; 492:153-168.
102. Fridovich T. The biology of oxygen radicals. Science 1978; 201:875-880.
103. Burton GW, Ingold KU. Autoxidation of biological molecules 1. The antioxidant activity of vitamin E and related chain breaking phenolic antioxidants in vitro. J. Am. Chem. Soc. 1981; 103:6472-6477.
104. Fantone JC, Ward PA. Role of oxygen-derived free radicals in leukocyte-dependent inflammatory reactions. Am. J. Path. 1982; 107(3):397-418.
105. Malech HL, Gallin JI. Neutrophils in human diseases. N. Engl. J. Med. 1987; 317:687-694.
106. Root RK, Metcalf JA. H_2O_2 release from human granulocytes during phagocytosis: relationship to superoxide anion formation and cellular catabolism of H_2O_2: studies with normal and cytochalasin B-treated cells. J. Clin. Invest. 1977; 60:1266-1279.
107. Roos DJ, Hamon-Müller WT, Weening RJ. Effect of cytochalasin-B on the oxidative metabolism of human peripheral blood granulocytes. Biochem. Biophys. Res. Commun. 1976; 68:43-50.
108. Patriarca P, Cramer R, Moncalvo S et al. Enzymatic basis of metabolic stimulation in leukocytes during phagocytosis: the role of activated NADPH oxidase. Arch. Biochem. Biophys. 1971; 145:255-262.
109. Babior BM, Curnutte JT, McMurrich BJ. The particulate superoxide-forming system from human neutrophils: properties of the system and further evidence supporting its participation in the respiratory burst. J. Clin. Invest. 1976; 58:989-996.
110. Test ST, Weiss SJ. Quantitative and temporal characterization of the extracellular H_2O_2 pool generated by human neutrophils. J. Biol. Chem. 1984; 259:399-405.
111. Haber F, Weiss J. The catalytic decompensation of hydrogen peroxide by iron salts. Proc. Roy. Soc. Lond. Series A 1934; 147:332-351.
112. Halliwell B. An attempt to demonstrate a reaction between superoxide and hydrogen peroxide. FEBS Lett. 1976; 72:8-10.
113. McCord JM, Day ED. Superoxide-dependent production of hydroxyl radical catalysed by an iron-EDTA complex. FEBS Lett. 1978; 86:139-142.
114. Aust SD, Svingen BA: The role of iron in enzymatic lipid peroxidation, in Pryor WA, (eds): Free Radicals in Biology. New York, Academic Press, Inc., 1982, pp 1-28.
115. Aust SD, Morehouse LA, Thomas CE. Role of metals in oxygen radical reactions. Free Radic Biol Med 1985; 1:3-25.
116. Halliwell B. Superoxide-dependent formation of hydroxyl radicals in the presence of iron chelates. Is it a mechanism for hydroxyl radical formation in biochemical systems? FEBS Lett. 1978; 92:321-326.
117. Korthuis RJ, Granger DN, Townsley MI et al. The role of oxygen-derived free radicals in ischemia-induced increases in canine skeletal muscle vascular permeability. Circ Res 1985; 57:599-609.
118. Hernandez LA, Grisham MB, Granger DM. A role for iron in oxidant-mediated ischemic injury to intestinal microvasculature. Am. J. Physiol. 1987; 253(Gastrointest. Liver Physiol.):G49-G53.
119. Smith JK, Carden DL, Grisham MB et al. Role of iron in postischemic microvascular injury. Am. J. Physiol. 1989; 256(Heart Circ. Physiol. 25):H1472-H1477.
120. Ambrosio G, Zweier JL, Jacobus WE et al. Improvement of postischemic myocardial function and metabolism induced by administration of deferoxamine at the time of reflow: The role of iron in the pathogenesis of reperfusion injury. Circulation 1987; 76:906-915.
121. Fantini GA, Yoshioka T. Deferoxamine prevents lipid peroxidation and attenuates reoxygenation injury in postischemic skeletal muscle. Am J Physiol 1993; 264(Heart. Circ Physiol. 33):H1953-H1959.
122. Ward PA, Till GO, Kunkel R et al. Evidence for role of hydroxyl radical in complement and neutrophil-dependent tissue injury. J. Clin. Invest. 1983; 72:789-801.

123. Varani J, Fligiel SEG, Till GO et al. Pulmonary endothelial cell killing by human neutrophils: possible involvement of hydroxyl radical. Lab. Invest. 1985; 53: 656-663.
124. Fligiel SEG, Ward PA, Johnson KJ et al. Evidence for a role of hydroxyl radical in immune-complex-induced vasculitis. Am. J. Pathol. 1984; 115:375-382.
125. Ambruso DR, Johnston RB Jr. Lactoferrin enhances hydroxyl radical production by human neutrophils, neutrophil particulate fractions, and an enzymatic generating system. J. Clin. Invest. 1981; 67:352-360.
126. Aruoma OI, Halliwell B. Superoxide-dependent and ascorbate-dependent formation of hydroxyl radicals from hydrogen peroxide in the presence of iron: are lactoferrin and transferrin promoters of hydroxyl radical generation? Biochem. J. 1987; 241:273-278.
127. Baldwin DA, Jenny ER, Aisen P. The effect of human serum transferrin and milk lactoferrin on hydroxyl radical formation from superoxide and hydrogen peroxide. J. Biol. Chem. 1984; 259:13391-13394.
128. Winterbourn CC. Lactoferrin-catalysed hydroxyl radical formation: additional requirement for a chelating agent. Biochem. J. 1983; 210:15-19.
129. Britigan BE, Rosen GM, Thompson BY et al. Stimulated human neutrophils limit iron-catalyzed hydroxyl radical formation as detected by spin-trapping techniques. J. Biol. Chem. 1986; 261:17026-17032.
130. Winterbourn CC. Myeloperoxidase as an effective inhibitor of hydroxyl radical production: implications for the oxidative reactions of neutrophils. J. Clin. Invest. 1986; 78:545-550.
131. Gannon DE, Varani J, Phan SH et al. Source of iron in neutrophil-mediated killing of endothelial cells. Lab. Invest. 1987; 57(1):37-44.
132. Cambria RA, Anderson RJ, Dikdan G et al. Leukocyte activation in ischemia-reperfusion injury of skeletal muscle. J. Surg. Res. 1991; 51:13-17.
133. Carden DL, Smith JK, Korthuis RJ. Oxidant-mediated CD_{18}-dependent microvascular dysfunction induced by complement-activated granulocytes. Am. J. Physiol. 1991; 260(Heart Circ. Physiol. 29): H1144-H1152.
134. Perez HD, Weksler BB, Goldstein IA. Generation of a chemotactic lipid from arachidonic acid by exposure to a superoxide-generating system. Inflammation 1980; 4:313-328.
135. Petrone WF, English DK, Wong K et al. Free radicals and Inflammation: superoxide-dependent activation of a neutrophil chemotactic factor in plasma. Proc. Natl. Acad. Sci. USA 1980; 77:1159-1163.
136. Del Maestro RF, Planker M, Arfors KE. Evidence for the participation of superoxide anion radical in altering the adhesive interaction between granulocytes and the endothelium. Int. J. Microcirc. Clin. Exp. 1982; 1:105-120.
137. Suzuki M, Inauen W, Kvietys PR et al. Superoxide mediates reperfusion-induced leukocyte-endothelial cell interactions. Am. J. Physiol. 1989; 257(Heart Circ. Physiol 26):H1740-H1745.
138. Test ST, Weiss SJ. The generation and utilization of chlorinated oxidants by human neutrophils. Adv. Free Radical Biol. Med. 1986; 2:91-116.
139. Johnson KJ, Fantone JCI. In vivo damage of rat lungs by oxygen metabolites. J. Clin. Invest. 1981; 67:983-993.
140. Johnson RJ, Couser WG, Chi EY et al. New mechanism for glomerular injury: myeloperoxidase-hydrogen peroxide-halide system. J. Clin. Invest. 1987; 79:1379-1387.
141. Vracko R. Basal lamina, scaffold-anatomy and significance for maintenance of orderly tissue structure. Am. J. Pathol. 1974; 77:314-346.
142. Hay ED, (ed): Cell Biology of the Extracellular Matrix. New York, Plenum Press, 1991, pp 468.
143. Janoff A. Elastase in tissue injury. Annu. Rev. Med. 1985; 36:207-216.
144. Henson PM, Henson JE, Fitlschen C et al.: Phagocytic cells, degranulation, and secretion, in Gallin JI, Goldstein IM, Snyderman R, (eds): Inflammation: Basic Principles and Clinical Correlates. New York, Raven Press, 1988, pp 363-380.
145. Travis J, Salvesen GS. Human plasma proteinase inhibitors. Annu. Rev. Biochem. 1983; 52:655-709.

146. Weiss SJ, Regiani S. Neutrophils degrade subendothelial matrices in the presence of α_1-proteinase inhibitor: cooperative use of lysosomal proteinases and oxygen metabolites. J. Clin. Invest. 1984; 73:1297-1303.
147. Kramps JA, VanTwisk C, Dijkman JH. Oxidative inactivation of bronchial antileukoprotease by triggered polymorphonuclear leukocytes. Am. Rev. Resp. Dis. 1987; 135:290 [abst.].
148. Mehta J, Dinerman J, Mehta P et al. Neutrophil function in ischemic heart disease. Circulation 1989; 79:549-556.
149. Bell D, Jackson M, Nicoll JJ et al. Inflammatory response, neutrophil activation, and free radical production after acute myocardial infarction: effect of thrombolytic treatment. Br. Heart J. 1990; 63:82-87.
150. Ranjadayalan K, Umachandran V, Davies SW et al. Thrombolytic treatment in acute myocardial infarction: neutrophil activation, peripheral leucocyte responses, and myocardial injury. Br. Heart J. 1991; 66:10-14.
151. Zimmerman BJ, Granger DN. Reperfusion-induced leukocyte infiltration: role of elastase. Am. J. Physiol. 1990; 259(Heart Circ. Physiol. 28):H390-394.
152. Inauen W, Granger DN, Meininger CJ et al. Anoxia-reoxygenation-induced, neutrophil-mediated endothelial cell injury: role of elastase. Am. J. Physiol. 1990; 259(Heart Circ. Physiol. 28):H925-H931.
153. Harlan JM, Killen PD, Harker LA et al. Neutrophil-mediated endothelial injury in vitro: mechanisms of cell detachment. J. Clin. Invest. 1981; 68:1394-1403.
154. Smedly LA, Tonnesen MG, Sandhaus RA et al. Neutrophil-mediated injury to endothelial cells: enhancement by endotoxin and essential role of neutrophil elastase. J. Clin. Invest. 1986; 77:1233-1243.
155. Weiss SJ, Peppin GJ. Collagenolytic metalloenzymes of the human neutrophil; characteristics, regulation, and potential function in vivo. Biochem. Pharmacol. 1986; 35:3189-3197.
156. Weiss SJ, Peppin G, Ortiz X et al. Oxidative autoactivation of latent collagenase by human neutrophils. Science 1985; 227: 747-749.
157. Peppin GJ, Weiss SJ. Activation of the endogenous metalloproteinase, gelatinase, by triggered human neutrophils. Proc. Natl. Acad. Sci. USA 1986; 83:4322-4326.
158. Harman JW. Significance of local vascular phenonena in the production of ischemic necrosis in skeletal muscle. Am. J. Path. 1948; 24:625-641.
159. Strock PE, Majno G. Vascular responses to experimental tourniquet ischemia. Surg. Gynecol. Obstet. 1969; 129:309-318.
160. Strock PE, Majno G. Microvascular changes in acutely ischemic rat muscle. Surg. Gynecol. Obstet. 1969; 129:1213-1224.
161. Schmid-Schönbein GW. Capillary plugging by granulocytes and the no-reflow phenomenon in the microcirculation. Fed. Proc. 1987; 46:2397-2401.
162. Braide M, Amundson B, Chien S et al. Quantitative studies on the influence of leukocytes on the vascular resistance in a skeletal muscle preparation. Microvasc. Res. 1984; 27:331-352.
163. Sutton DW, Schmid-Schönbein GW. Elevation of organ resistance due to leukocyte perfusion. Am. J. Physiol. 1992; 262(Heart Circ. Physiol. 31):H1646-H1650.
164. Harris AG, Skalak TC. Effects of leukocyte activation on capillary hemodynamics in skeletal muscle. Am. J. Physiol. 1993; 264(Heart. Circ. Physiol. 33):H909-H916.
165. Del Zoppo GJ, Schmid-Schönbein GW, Mori E et al. Polymorphonuclear leukocytes occlude capillaries following middle cerebral artery occlusion and reperfusion in baboons. Stroke 1991; 22:1276-1283.
166. Engler RL, Dahlgren MD, Morris DD et al. Role of leukocytes in response to acute myocardial ischemia and reflow in dogs. Am. J. Physiol. 1986; 251(Heart Circ. Physiol. 20):H314-H322.

CHAPTER 4

Platelet-Activating Factor

Peter J. Pappas

Walter N. Durán

Platelet activating factor (PAF) is a potent mediator of inflammation. For the purposes of this discussion, inflammation can be defined as the reaction of vascularized tissues to local injury. Typically, the response to tissue injury involves transient arteriolar vasonconstriction, followed by vasodilation of the local microvascular bed and arterioles. The increased blood flow caused by the vasodilation is associated with a concomitant increase in microvascular permeability. The increased permeability leads to slowing of the microcirculation due to exudation of protein rich fluid into the extracellular space. Subsequently, leukocyte margination, adherence and emigration occur.

PAF has been reported to influence every aspect of the above described inflammatory response. When administered systemically, PAF causes hypotension, increased vascular permeability, hemoconcentration, pulmonary hypertension, bronchoconstriction, thrombocytopenia, endotoxemia and neutropenia in various animal models.[1] Locally, PAF increases vascular permeability and results in the accumulation of polymorphonuclear leukocytes (PMNs) and monocytes in man.[2,3] Even though PAF is produced by a variety of cell types in man, including PMNs, monocytes, macrophages, basophils, endothelial cells and platelets, it is still not clear what role PAF plays in pathologic or physiologic circumstances.[4-11] This chapter will focus on the effects of PAF on the microcirculation and its role in the ischemia-reperfusion (I/R) syndrome in skeletal muscle.

HISTORY

PAF was discovered in 1966 when Barbaro et al reported that rabbit platelets released histamine after antigen presentation[12] and noted that this

Ischemia-Reperfusion Injury of Skeletal Muscle, edited by Gary A. Fantini, M.D.;

phenomenon was leukocyte dependent.[13] Benveniste et al in 1972 proposed that a soluble factor, which they called platelet activating factor, was released from leukocytes, and caused the histamine release observed by Barbaro et al.[14] Between 1972 and 1977, Benveniste et al described the characteristics and molecular structure of PAF.[15,16] In addition, the molecular structure of PAF was described concurrently by Demopoulos and colleagues in 1979.[18] Chemically PAF is 1-alkyl-2(R)-acetyl-glycero-3-phosphorylcholine.[17,18] (Fig. 4.1).

Synthetic PAF subsequently became available in 1980.[19] Since that time, numerous chemical analogs with alterations at the C1, C2, or C3 positions have been synthesized. Experiments performed with these analogs have indicated that changes in stereochemical configuration are associated with significant reduction in potency.[20] Due to PAF's stereochemical specificity and numerous ligand binding studies, the existence of specific PAF receptors was recognized.[1] These receptors were subsequently determined to initiate and mediate several intracellular and extracellular signalling mechanisms responsible for PAF's physiologic actions. In 1982, Lapetina[51] reported that PAF receptor binding stimulated phosphoinositide turnover. It is now known that inositol phospholipids are metabolized by phospholipases which generate inositol triphosphate (IP3) diacylglycerol (DG) and arachidonic acid (AA). IP3 and DG serve as intracellular second messengers and play a role in intracellular Ca^{+2} mobilization.[21,22] AA is used by the cyclooxygenase and lipoxigenase systems to produce prostaglandins and leukotrienes.

The mechanistic interrelationship between PAF receptor binding and phosphoinositide turnover was concurrently reported by Hwang et al[23] and Barzaghi in 1986.[24] They reported that the effects mediated by PAF receptor binding were coupled to G-protein linked inhibition of adenylate cyclase and activation of phospholipase C respectively.[23,24] Confirmation of this observation came from Honda et al in 1991 when this group cloned a PAF receptor from guinea pig lung and reported the receptor to be homologous with other G-protein associated receptors.[25] The G-protein itself, however, has yet to be isolated.

BIOSYNTHESIS

PAF is not known to be stored in any cell.[26] PAF is present in the form of the inactive precursor 1-O-alkyl-2-sn-glycero-3-phosphocholine (Alkylacyl-GPC) which is linked to membrane phospholipid structures.[1] PAF is synthesized by either of two pathways depending on cell type (Fig. 4.2). The first pathway is known as the deacylation-reacylation cycle or the remodeling pathway.[26,1] The first step in PAF biosynthesis

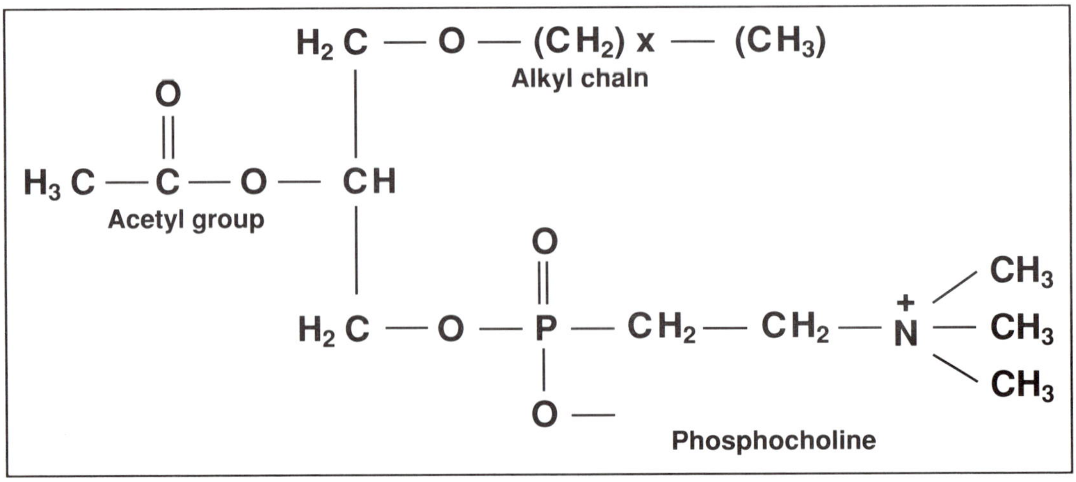

Fig. 4.1. Chemical structure of platelet activating factor.

involves the activation of membrane bound phospholipase A_2 (PLA_2). PLA_2 then catalyzes the hydrolysis of an esterified fatty acyl residue in the *sn*-2 position of a putative precursor molecule, 1-O-alkyl-2-*sn*-glycero-3-phosphocholine.[26,27] This hydrolysis yields an intermediate, 1-O-alkyl-*sn*-glycero-3-phosphocholine (lyso-PAF) and a free fatty acid which is usually arachidonic acid. Phospholipase A_2 prefers a phospholipid that has an AA at the 2-position.[26] The AA is then available as a precursor for the lipoxygenase and cyclooxygenase systems.[26,27] Next, lyso-PAF is acetylated by acetyltransferase, a membrane associated enzyme complex, and acetyl-CoA to form PAF. This enzyme is activated by phosphorylation and calcium influx. The kinase responsible for this process is either an unknown cyclic-adenosine monophosphate (c-AMP)-dependent protein kinase or protein kinase C (PKC).[26,27,1] PAF is degraded by conversion to lyso-PAF by an acetylhydrolase which is present in the intracellular and extracellular compartments.[1]

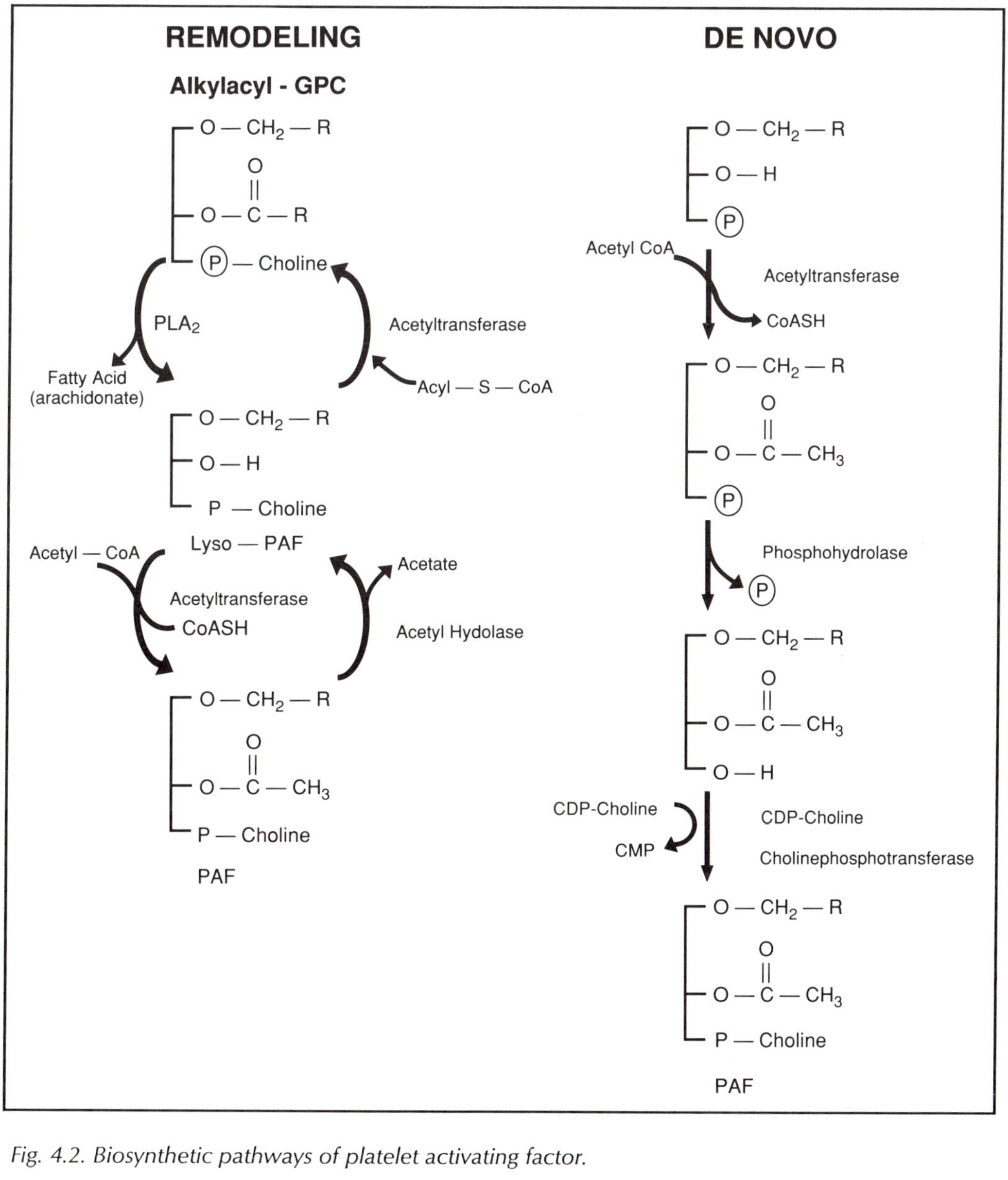

Fig. 4.2. Biosynthetic pathways of platelet activating factor.

Acetylhydrolase removes the acetyl group from the 2(R) position of PAF forming lyso-PAF. Lyso-PAF is then acylated at the 2(R) position, usually by an arachidonic acid residue from phosphatidylcholine, by acyltransferase. This forms the PAF precursor alkylacyl-GPC, which becomes integrated back into the cell membrane.[1] Acyltransferase is inhibited by calcium. Therefore, during cell activation, PAF production is favored.[1]

A second route of PAF synthesis, termed the de novo pathway, involves the transfer of a phosphocholine group into 1-alkyl-2(R)-acetyl-glycerol by a specific CDP-choline cholinephosphotransferase. The CDP choline-phosphotransferase is different from that involved in the conversion of diacylglycerol into phosphatidylcholine (PC).[1,26,27] Some cells do not produce PAF by the de novo pathway and its physiologic role remains unclear.[1,26]

MECHANISMS OF SIGNAL TRANSDUCTION

PAF's effects are mediated through binding to specific PAF receptors on the surface of target cells. Radioligand binding studies using [3H]PAF and structurally different PAF antagonists have provided the characteristics of the PAF receptor and its heterogeneity.[28] For example, Hwang reported that various ions can alter the binding properties of PAF receptors and that these receptors exist in high and low affinity states. These findings suggest that PAF receptors exist in different conformational states.[29]

There is also strong evidence that PAF receptors mediate their actions via membrane associated G proteins. In 1991, Honda et al cloned a PAF receptor using a ligand-specific cloning strategy that involved gene expression in *Xenopus laevis* oocytes. The receptor showed the familiar seven hydrophobic transmembrane segments that have become the characteristic hallmark of G protein-coupled receptors.[28,30] The cytoplasmic tail of the residue contained four serine, five threonine, and two tyrosine residues, all of which may be involved in protein kinase activation. Several asparagine residues were present at the external surface of the receptor and could serve as sites for attachment of glycosylated residues.[28] Further support for the receptor being coupled to G proteins is the fact that PAF stimulates GTPase activity, pertussis toxin inhibits PAF-stimulated responses in many systems suggesting the presence of a G inhibitory-associated protein, and GTP causes a shift in the binding of PAF.[28]

Most studies to date emphasize PAF's extracellular release and target cell effect. However, Henson has reported that most of the PAF synthesized by neutrophils is retained intracellularly.[27] In neutrophils, the proportion of PAF excreted after stimulation never exceeded 30% of the total synthesized. This fraction appears to be even lower in endothelial cells.[27] This observation is not inconsistent with PAF's extracellular effects. PAF is so potent that only small amounts are needed in order to mediate its inflammatory effects.[27]

SECOND MESSENGER PRODUCTION MEDIATED BY PHOSPHOLIPID TURNOVER

PAF's binding to its receptor causes the generation of intracellular second messengers via phospholipid metabolism. PAF has been reported to activate phospholipase A_2, C and D in various cells. PLA_2 cleaves membrane bound phospholipids at the 2(R) position producing a fatty acid and lyso-PAF. The fatty acid most often produced is AA. Arachidonic acid can be utilized by the cylo-oxygenase and lipoxygenase systems where conversion into active prostaglandins and/or leukotrienes can occur.[31,28] Similarly, phospholipase D (PLD) acts upon membrane bound phospholipids at the 2(R) position to produce phosphatidic acid (PA). PA can be metabolized by phosphatidic acid phosphohydrolases to produce diacylglycerol (DAG) which can activate PKC or be converted into AA.[28]

PAF's main signaling mechanism appears to be mediated through phospholipase C (PLC) turnover of phosphotidylinositides (PPI). PAF receptor binding causes activation of PLC and inhibition of adenylate cyclase via G-protein coupling. Upon receptor activation, GTP binds to the α subunit of either a stimulatory (G_s) or inhibitory (G_i) G protein, causing dissociation from its β and γ subunits.

The now functionally dissociated α subunit cleaves GTP, activates PLC and inhibits adenylate cyclase. Activated PLC then converts PPI into inositol trisphosphate (IP_3) and DAG. IP_3 causes an increase in intracellular calcium by mobilization of intracellular storage sites. The rough endoplasmic reticulum is thought to be the storage site acted upon by IP_3.[31] Extracellular calcium mobilization is hypothesized to be mediated via PAF receptor linked calcium channels.[21] However, data regarding the effects of calcium channel blockers on PAF stimulated intracellular calcium changes has yielded conflicting results, thereby casting doubts on the existence of these channels.[21] Once mobilized, calcium and DAG then serve to activate PKC.[31]

PROTEIN KINASE C AND TYROSINE KINASE

PAF receptor binding inhibits adenylate cyclase and decreases intracellular cyclic adenosine monophosphate (cAMP). Cyclic-AMP causes inhibition of PPI turnover and intracellular calcium elevation.[21] Activation of PKC similarly inhibits PPI turnover and is negatively coupled to PLC stimulation.[21] In neutrophils, activated PKC downregulates high affinity PAF receptors which correlates with a decrease in intracellular calcium.[21] Staurosporine, a putative PKC inhibitor, inhibited PKC-mediated protein phosphorylation and potentiated PAF stimulation of IP3 production.[21] These data imply a negative feedback role of PKC on PAF stimulated PPI turnover.[21] If so, PKC may be responsible for the desensitization of cells to prolonged PAF stimulation. However, there is one study which showed that staurosporine failed to prevent PLC desensitization after PAF application.[32] The exact role of PKC after PAF stimulation and the exact proteins phosphorylated by PKC are not known. Although the proteins are unknown, PKC does have a broad specificity which differs from cyclic nucleotide dependent and calcium/calmodulin-dependent protein kinases. Thus, this enzyme has been reported to phosphorylate histone H1, troponin T, smooth muscle myosin light chain, myosin light chain kinase, tyrosine hydroxylase, and IP_3 phosphatase.[31]

Tyrosine kinase (TK) has also been linked to PLC activation.[21] PAF has been reported to be responsible for the phosphorylation of a 50Kd and 60Kd protein. Both these proteins reacted with a monoclonal antibody directed towards phosphotyrosine residues. Genistein, a TK inhibitor, blocked the phosphorolation of these two proteins and inhibited binding by the phosphortyrosine specific monoclonal antibody after PAF stimulation. Genistein has also been reported to inhibit PAF stimulated platelet aggregation and PPI turnover.[21] The mechanistic association between PAF and TK is unknown. The recognition of the numerous tyrosine residues present in PAF receptors and the above findings raise the question of whether the PAF receptor is a substrate for TK or the actual enzyme itself.[1] Numerous experiments however, have suggested that TK and PKC activity are involved in the induction of early response genes and that the expression of these genes are PAF receptor dependent.[28] The exact mechanism responsible for the upregulation of these genes remains to be elucidated.

SUMMARY OF SIGNAL TRANSDUCTION

PAF binds to specific PAF receptors on various target cells. Receptor binding activates a Gs which stimulates PLC, PLA_2, and PLD and/or a Gi protein which inhibits adenylate cyclase. The concurrent decrease in intracellular cAMP and the G-protein stimulation of PLC causes the breakdown of PPI into IP_3 and DAG. IP_3 mobilizes intracellular calcium stores from the rough endoplasmic reticulum. Further increases in intracellular calcium occurs through recruitment of extracellular calcium, presumably via as yet unidentified PAF linked calcium channels. Calcium and DAG then activate PKC whose exact function is not known. PKC may serve to inactivate PLC and along with TK may be responsible for the upregulation of early response genes. Stimulation of PLA_2 and PLD produces AA and PA. PA can be converted to AA which then becomes available for use in cyclooxygenase and lipoxygenase systems. These mediators can then be released into the extracellular space and serve to regulate an inflammatory response (Fig. 4.3).

CELLULAR RESPONSES AND EFFECTS OF CYTOKINES

Platelets and Eosinophils

The name PAF is somewhat of a misnomer because this lipid autacoid is now known to have effects on numerous cell types. Platelets lack significant amounts of the enzyme acetyltransferase, which is essential in the conversion of lyso-PAF to active PAF. Since platelets can only produce large amounts of lyso-PAF, it is hypothesized that platelets serve as donors to leukocytes which synthesize PAF. This is consistent with Barbaro's original observation in 1966 that platelet release of histamine after antigenic challenge was leukocyte dependent.[12] Once produced, PAF has been reported to be a powerful chemoattractant of platelets and eosinophils in the airways of various animal models. In guinea pigs, PAF causes the release of platelet factor 4 (PF4) and platelet derived growth factor (PDGF), both of which are extremely potent agonists of eosinophil chemotaxis and degranulation.[33] Due to this association, it is felt that PAF may play a role in the pathophysiology of asthma in humans.

Polymorphonuclear Leukocytes

PAF and lyso-PAF are released from stimulated human PMNs. PAF induces leukocyte activation and causes in vivo leukopenia and margination. In vitro activation leads to lysosomal degranulation and formation of superoxide and arachidonate metabolites. PAF also primes PMN responses to various agonists. "Priming" may be defined as the amplification of a cellular response to a stimulus by prior exposure of the cell to various

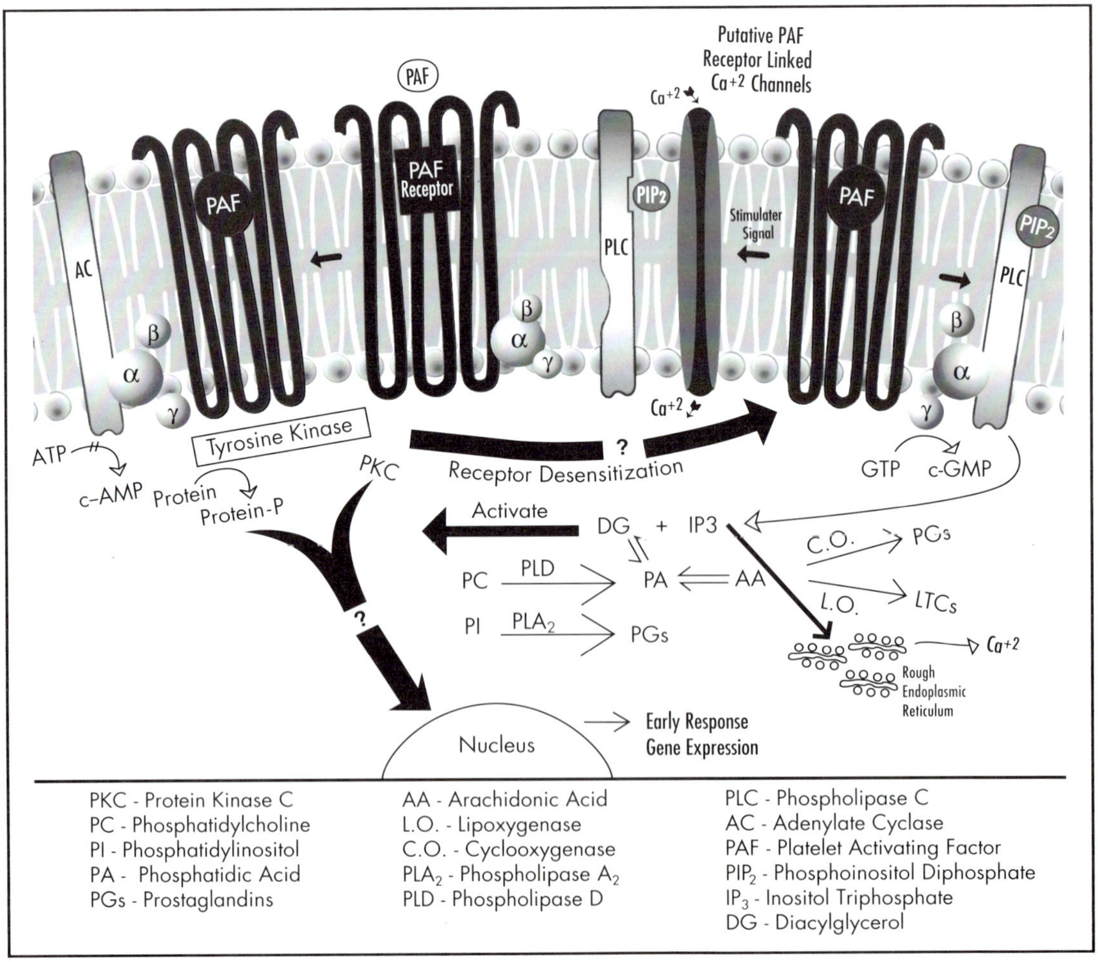

Fig. 4.3. Signal transduction pathways for platelet activating factor.

agonists. Primed responses include: (1) increased respiratory burst as evidenced by enhanced superoxide release, (2) degranulation as measured by elastase release, and (3) increased adhesion to endothelial cells and increased microvascular permeability.[33,37] PAF primes or enhances the response of PMNs to thrombin, N-formylmethionylleucyl phenylalanine (FMLP), granulocyte-monocyte colony stimulating factor (GM-CSF) and tumor necrosis factor (TNF). GM-CSF primes PAF induced arachidonate release and A23187 calcium ionophore evoked leukotriene B4 (LTB4) production. GM-CSF priming is thought to be mediated via G proteins and stimulation of uridine incorporation into m-RNA. TNF can prime PMNs either alone or with PAF. After exposure to TNF, PMNs show an increased respiratory burst, activation of PLA_2, and increased release of cathepsin G.[34]

Endothelial Cells

PAF causes cultured endothelial cells to change their shape, retract and lose reciprocal contact. This effect of PAF is thought to be due to a decrease in the contact protein vinculin, which is observed in areas associated with adhesion plaques.[33] PAF also promotes albumin diffusion between cultured endothelial cells grown on fibronectin-coated polycarbonate filters, which restrict I^{135} albumin diffusion.[33] I^{135} albumin diffusion is observed at concentrations similar to that required for PAF induced endothelial cell cytoskeleton alterations. In vivo endothelial cell retraction in the mesenteric beds of guinea pigs led to collagen exposure, platelet aggregation, PMN recruitment and margination.[33] These phenomena coincide with PAF's ability to increase permeability in the microcirculation.

Rapid production of PAF by endothelial cells is observed after stimulation with leukotrienes, histamine, ATP, and bradykinin. PAF in endothelial cells is produced by either the remodeling pathway or the de novo pathway. The remodelling pathway, however, is cytokine sensitive and can be upregulated by exposure to TNF and/or IL-1. The effects of these two cytokines are additive and not synergistic, suggesting a common signaling pathway. Increased activity of the acetyltransferase enzyme system after TNF and IL-1 exposure suggests that these cytokines are promoting transcription of mRNA involved in increasing the level or activity of the enzyme system.[33] Supporting this hypothesis is the fact that actinomycin D and cycloheximide inhibit the increased activity of TNF treated endothelial cells. The de novo pathway in ECs is not cytokine sensitive.

Unlike other cells, endothelial cell acetyltransferase is not only regulated by an unknown phosphorylation dephosphorylation enzyme but is also regulated by a Na^+/H^+ antiport system. Five-(n-ethyl-N-isopropyl)-amaloride, a Na^+/H^+ antiport inhibitor, elicits a dose-dependent inhibition of thrombin-mediated PAF production. The Na^+/H^+ antiport system is involved in regulating the cytoplasmic concentration of calcium, which is needed for activation of acetyltransferase.

Monocytes

Monocytes are known sources of TNFα and IL-1α and β.[34] TNFα is a powerful immunomodulatory cytokine that plays a role in endotoxic shock, macrophage activation and production of IL-1, GM-CSF and prostaglandins. It also modulates γ-interferon synthesis. When stimulated by PAF, monocyte secretion of these cytokines is amplified. PAF also amplifies the γ-interferon or lipopolysaccharide-induced production of TNF.

MICROCIRCULATORY EFFECTS OF PAF

The in vivo effects of PAF on the microcirculation have been studied extensively in our laboratory using a hamster cheek pouch model. Our studies have indicated that PAF causes the following responses in the microcirculation: (1) a dose-related vasoconstriction that is mediated by thromboxane $A_2(TXA_2)$ (2) a dose-related extravasation of macromolecules, (3) stimulation of leukocyte adhesion in an inverse dose response relationship, (4) vasoconstrictor and permeability effects appear to be mediated by different receptors in the precapillary arterioles and postcapillary venules, and (5) PAF can act as a primer for the microcirculatory responses elicited by histamine.[35-37]

PAF is a potent vasoconstrictor when applied topically (Fig. 4.4). The onset of the vasoconstrictor response produced is very rapid, reaching a maximum value usually within 2 minutes following topical application. Similarly, the duration of action may be prolonged, depending on the dose administered. In general, smaller arterioles respond more strongly than larger ones to a given PAF challenge. The vasoconstrictor response is mediated through PAF-induced production of TxA_2.[35] We have observed that dexamethasone, a PLA_2 inhibitor, and OKY-046, an inhibitor of TxA_2 synthase, completely abolished PAF induced arteriolar vasoconstriction. Kadsurenone, a PAF-receptor blocking agent, similarly completely abolished PAF induced arteriolar vasoconstriction. This observation implies a role for the cyclooxygenase system, thromboxane synthase and the PAF-specific receptor in mediating PAF induced vasoconstriction (Fig. 4.5). The lipoxygenase system was not evaluated, due to a lack of specific enzyme inhibitors for this system.

PAF induces extravasation of plasma and proteins from the microcirculation by increasing microvascular permeability (Fig. 4.6). Using intravital microscopy and fluorescently labelled macromolecules with fluorescein isothiocyanate conjugated to dextran (FITC-dx 150), we reported that topically applied PAF causes extravasation of FITC-dx 150 into the interstitial compartment of the hamster cheek pouch. The microvascular permeability response exhibited a dose response relationship, with the peak occurring at PAF 10^{-7} M (Fig. 4.7; reference 35). Using various PAF antagonists and prostaglandin inhibitors, we reported that the permeability response is mediated partially by PAF receptors and may potentially involve pathways related to leukotriene production (Fig. 4.8; ref. 35).

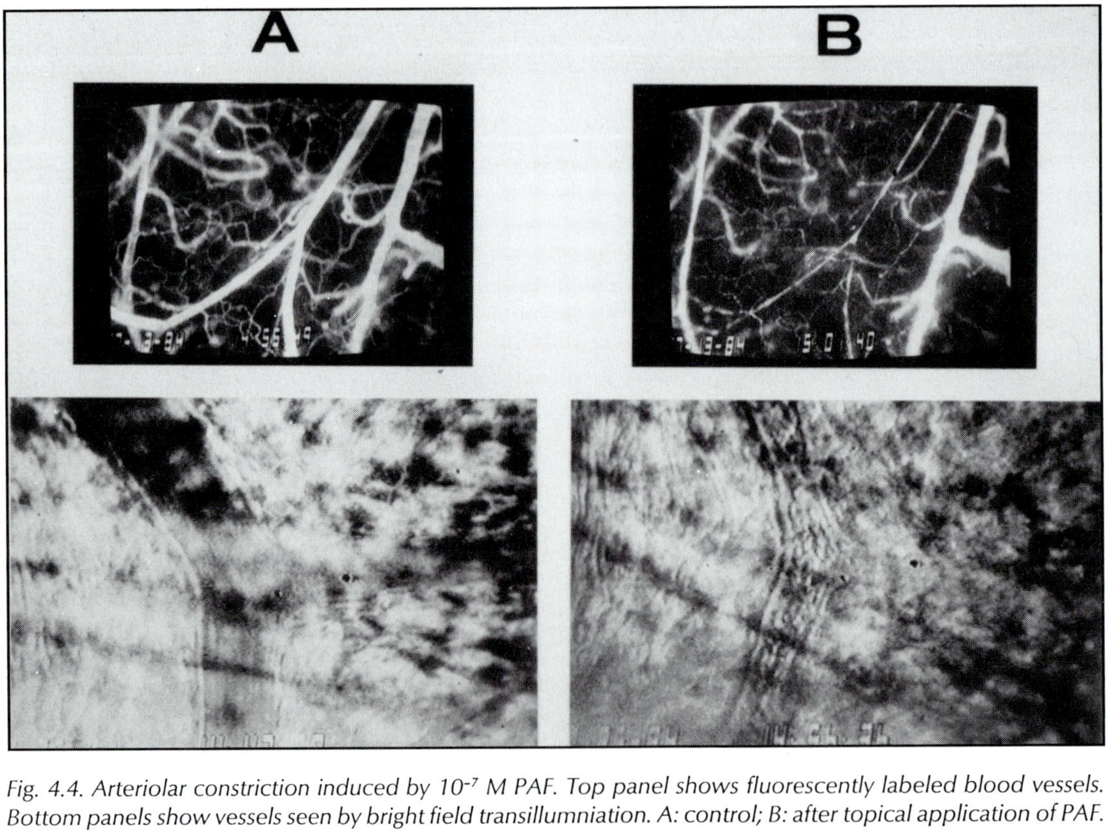

Fig. 4.4. Arteriolar constriction induced by 10^{-7} M PAF. Top panel shows fluorescently labeled blood vessels. Bottom panels show vessels seen by bright field transillumniation. A: control; B: after topical application of PAF.

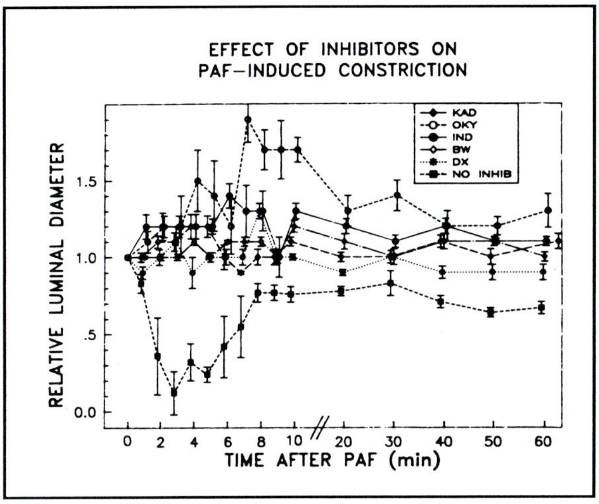

Fig. 4.5. Effect of pharmacological antagonists on PAF-induced arteriolar constriction. PAF 10^{-7} M was applied and arterioles with luminal diameters between 21 and 40 μm observed. KAD, kadsurenone; OKY, OKY-046; IND, indomethacin; BW, BW755C; DX, dexamethasone. Data shown as mean ± SD.

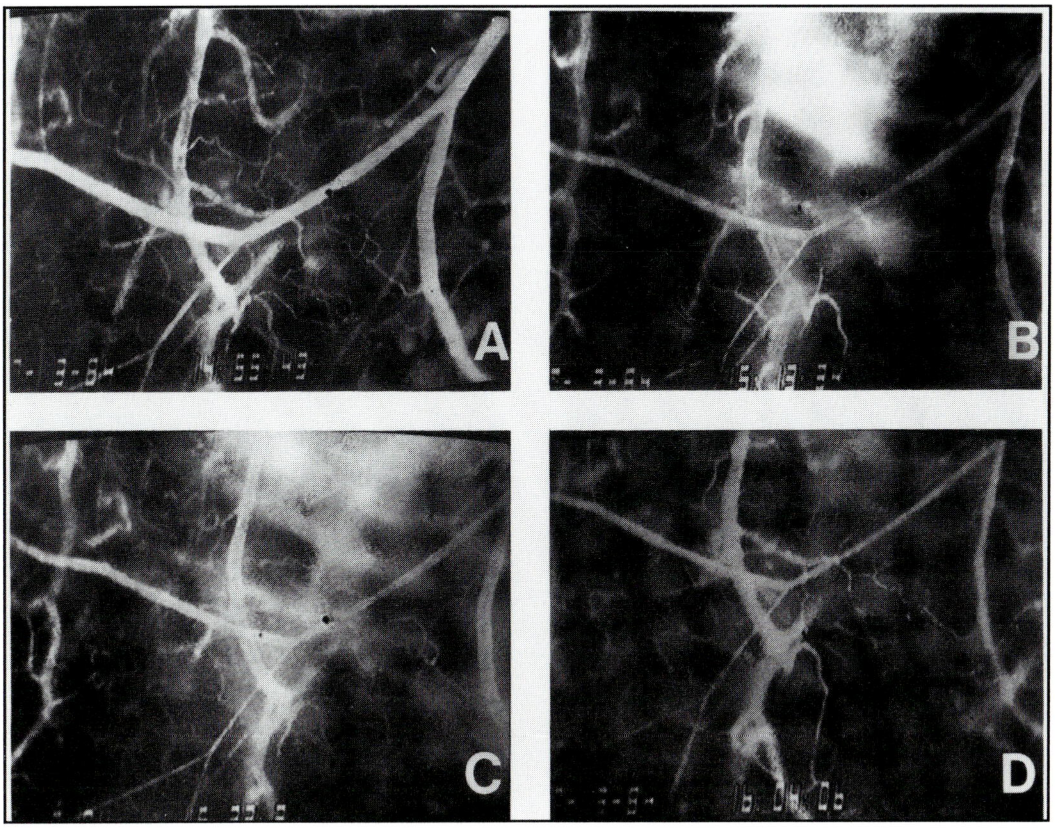

Fig. 4.6. Examples of vasoconstriction and macromolecular leakage induced by topical application of PAF 10^{-7} M in the hamster cheek pouch. FITC-dx 150 was injected intravenously 30 min before experimental interventions. (A) Control. Note absence of FITC-dx 150 in interstitial space. (B) Arteriolar vasoconstriction and leakage of FITC-dx at postcapillary venules are present 3 min after application of PAF. (C) Strong vasoconstriction and more generalized leakage 23 min after PAF application. (D) Arterioles are still constricted and diffuse leakage can be detected 54 min after PAF application.

PAF stimulates leukocytes to adhere to and migrate through microvascular walls in vivo (Fig. 4.9). A surprising finding, however, is that PAF-mediated leukocyte adhesion occurs in an inverse dose-response relationship to that required for postcapillary venular endothelium effects. The greatest effect on leukocyte adhesion was obtained at a PAF concentration of 10^{-9} M that neither produced a vasoconstrictor effect or altered microvascular permeability.[40]

PAF induced permeability and vasoconstrictor effects in vivo seem to be mediated via several types of receptors. Through a series of experiments, we have reported that several PAF antagonists abolished PAF induced arteriolar vasoconstriction, but selectively inhibited the permeability response (Ref. 36; Fig. 4.10). These differences may be due to the presence of structurally different PAF receptors and/or differential sensitivity/affinity between different microvascular segments.

As discussed earlier, priming may be defined as the amplification of a cellular response

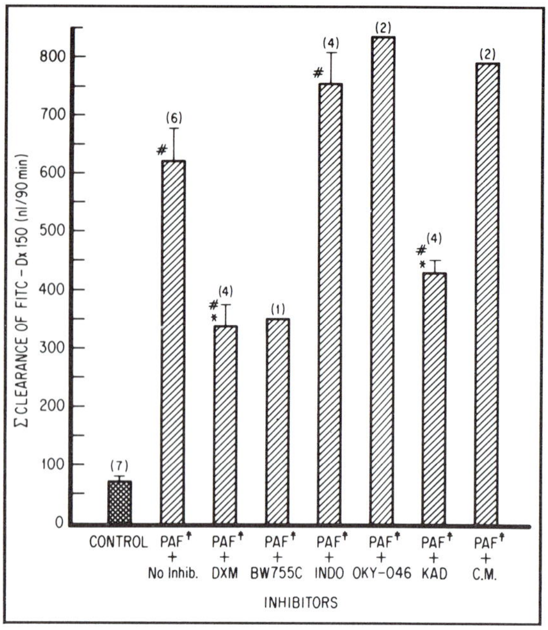

*Fig. 4.8. Effect of pharmacological antagonists on PAF-induced increased clearance of FITC-dx 150. *$p<0.05$ when compared with 10^{-7}M PAF. #$p<0.05$ when compared with control. DXM, dexamethasone; INDO, indomethacin; KAD, kadsurenone; C.M., chlorpheniramine maleate.*

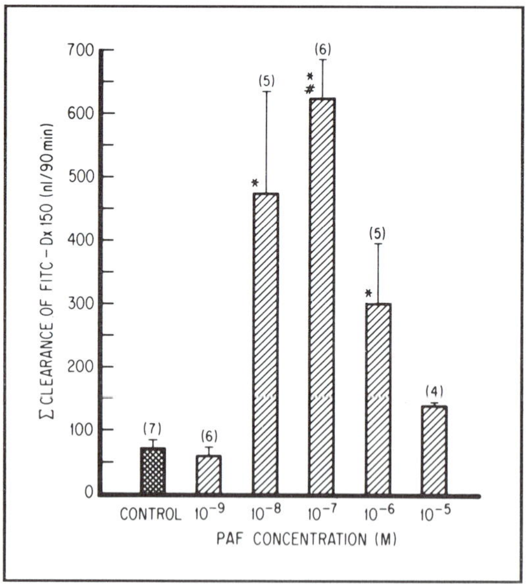

*Fig. 4.7. Dose-response histogram characterizing relation between PAF concentration and microvascular permselectivity. *$p<0.05$ compared with control. #$p<0.05$ compared with 10^{-9}, 10^{-6}, and 10^{-5} M PAF. Numbers in parentheses, number of experimental animals.*

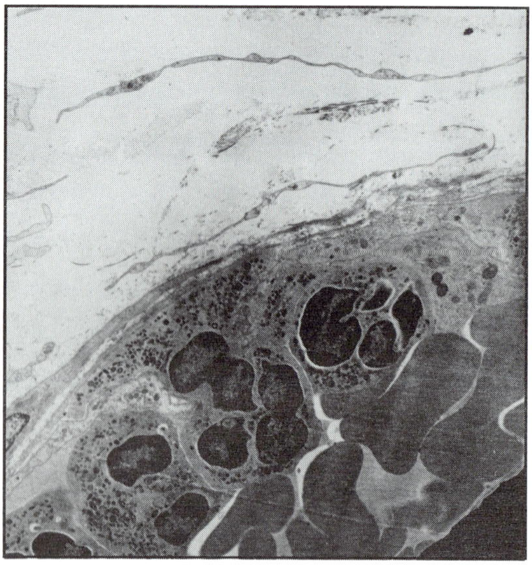

Fig. 4.9. Leukocyte adherence induced by topical application of 10^{-9} M PAF.

to a given stimulus due to the prior exposure of that cell to another agonist (Fig. 4.11). Priming should not be confused with synergism, which refers to the action of two agonists applied simultaneously, resulting in a response greater than the sum of the agents. PAF and histamine independently increase microvascular permeability to macromolecules in a dose dependent manner. The efficacious dose to achieve a change in permeability for PAF

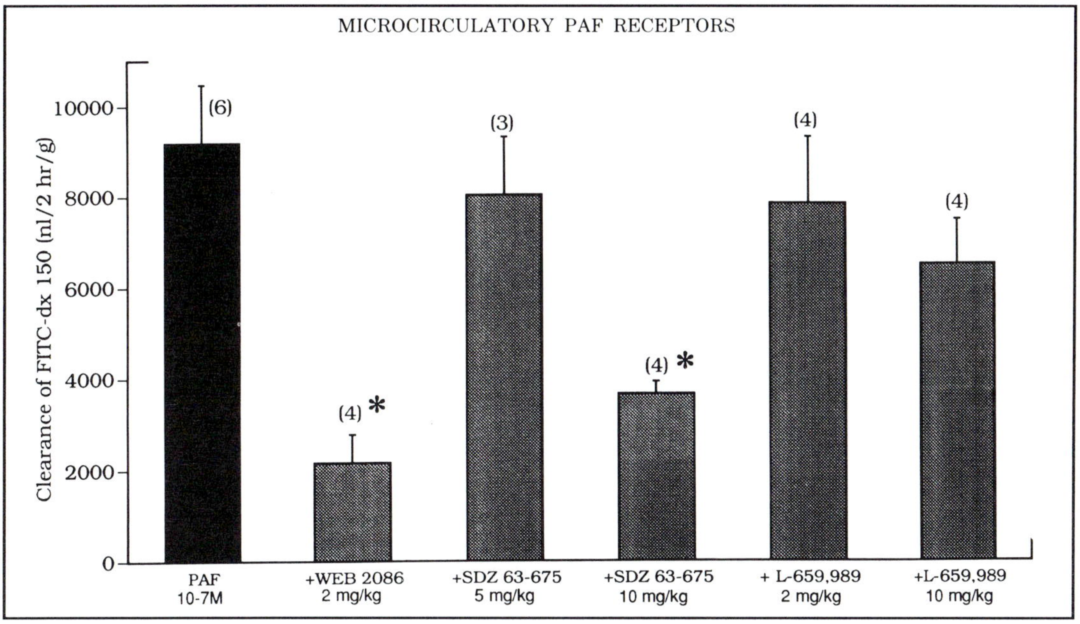

*Fig. 4.10. Pharmacological inhibition of PAF-induced microvascular permeability. Efficacy of 3 PAF receptor antagonists at reducing the rise in FITC-Dx 150 clearance elicited by topical challenge with 10^{-7} M PAF was tested. Data are presented as means ± SE. Numbers in parentheses show number of animals. *p<0.05 compared with 10^{-7} M PAF.*

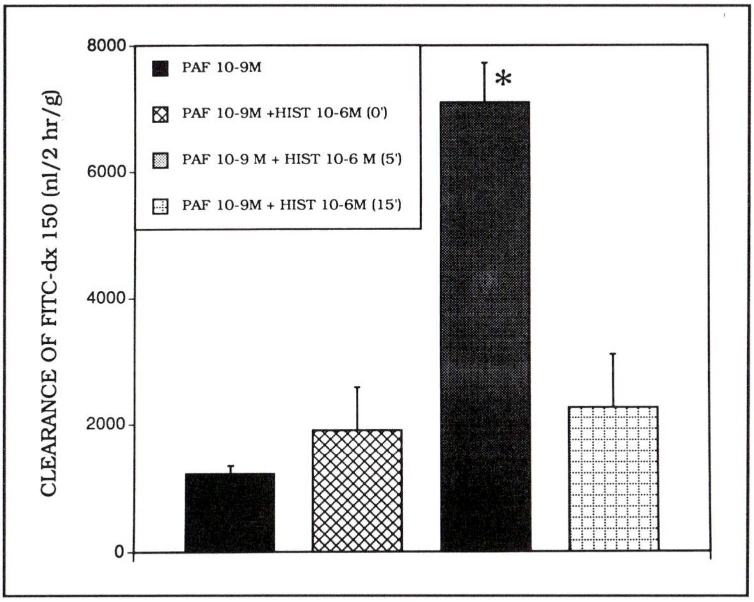

*Fig. 4.11. Effect of time intervals between agonists on the primed clearance response. The first bar, included as a reference, represents the effect of 10^{-9} M PAF on macromolecular clearance; this effect is not significantly different from baseline. The coapplication of PAF and histamine (time interval = 0) did not produce any significant rise in macromolecular clearance. Similarly, a 15 min interval (last bar) did not increase FITC-dx 150 clearance. Effective priming was obtained with a 5 min interval. Data are mean ± SEM (n = 6 per treatment group). *, P< 0.05 compared with 10^{-9} M PAF.*

in the hamster cheek pouch is 10^{-8} M and 10^{-5} M for histamine. We have reported that subthreshold doses of these agents are able to enhance the permeability modulating actions of each other.[37] We found that the priming response was receptor-mediated and that the dose of PAF used caused a significant increase in leukocyte adhesion to postcapillary venular endothelium in the hamster cheek pouch. What role the leukocyte plays in the priming response, if any, is not known. We also observed that leukotrienes play a role in the priming response, as the leukotriene inhibitor RG 5901 attenuated the macromolecular clearance response.

EFFECTS OF PAF DURING ISCHEMIA/REPERFUSION

A crucial event in the acute and chronic inflammatory response is leukocyte recognition and adhesion to target endothelial cells. Leukocyte-endothelial cell recognition is an active process requiring three sequential events: reversible adhesion, leukocyte activation and activation-dependent binding.[38] L-selectins (glycosylated carbohydrates on the surface of circulating neutrophils) mediate neutrophil "rolling" and are responsible for the reversible adhesion process. L-selectins sample the postcapillary venule endothelial environment for evidence of inflammation by searching for E-selectins, especially ELAM-1, on the surface of endothelial cells.[20] If recognition and binding occur, neutrophils become activated. Activated neutrophils shed their L-selectins and begin expressing the activation-dependent adhesion molecule Mac-1 (CD11/CD18 complex). Mac-1's counterligand is ICAM-1, which is expressed only on activated endothelial cells. If Mac-1 and ICAM-1 are present, activation dependent binding and neutrophil degranulation occur. However, not all the counterreceptors are known. Table 4.1 lists several known ligands and counterligands.

Rolling slows neutrophil transit through the inflamed postcapillary venule and allows it to sample the local environment for activating signals. Neutrophils activated by chemoattractants shed their L-selectins and upregulate the expression of β_2 integrins. Activators of neutrophils causing upregulation of β_2 integrins include PAF and TNFα. Upregulation of β_2 integrins on the neutrophil cell surface occurs in minutes. The integrins involved are the CD11a/CD18, CD11b/CD18, and CD11c/CD18 receptors. PMNs store CD11b and CD11c in cytoplasmic granules. Upon stimulation, the granules fuse with the cell membrane and increase the number of CD11b/CD18 and CD11c/CD18 receptors available for adhesion. Activation of the endothelial cells by inflammatory mediators causes upregulation of ICAM-1, which is the counterligand for the CD11/CD18 complex.

Using a hamster cheek pouch model,[40] we investigated the affects of PAF on leukocyte adhesion in the microcirculation. We found that: (1) PAF (10^{-11}M-10^{-7}M) increases the number of white cells adhering to postcapillary venular walls in an inverse dose-response manner, (2) peak number of adherent cells is obtained within the first 15 min following topical PAF application and remains unchanged for at least 3 hours, (3) the number of adhering white cells is dependent

Table 4.1. Leukocyte—endothelium ligands and counterligands that may play a role in ischemia/reperfusion injury

Primary Adhesion Pathways	
Leukocyte	Endothelial Cell
L-Selectin	High endothelial venule
CLA	E-selectin
Sialyl Lewis x	E-selectin
Sialyl Lewis y	P-selectin
Activation Dependent Adhesion Pathways	
Leukocyte	Endothelial Cells
LFA-1	ICAM 1 and 2
MAC-1	ICAM 1 and others
p150,95	?
VLA-4	VCAM-1

Modified from Springer TA. Nature 1990; 346:3425-3434 (ref. 39)

upon postcapillary venule size, and (4) PAF challenge does not increase the number of leukocytes adhering to arteriolar walls or the number of rolling leukocytes in venules and arterioles.[40] Previously, we had reported that PAF concentrations higher than 10^{-9}M induced arteriolar vasoconstriction, decreased blood flow, and increased clearance of FITC-dx 150, while 10^{-9}M and lower concentrations were without effect on these parameters. These experiments thus indicate that PAF appears to be a much more potent modulator of leukocyte adhesion than of arteriolar tone or macromolecular extravasation.[40-42] From these studies, it was apparent that PAF modulates leukocyte adhesion, arteriolar tone and permselectivity.

Other investigators have also reported that PAF may play a role in modulating the microcirculatory dysfunction observed in the I/R syndrome.[44] Continuing with the hamster cheek pouch model,[43] we investigated the role of PAF in modulating leukocyte-endothelial interactions and its effects on I/R. We measured the changes in the number of adherent leukocytes to postcapillary venular endothelium (per 100 µm vessel length) during I/R in the absence and presence of a specific PAF receptor antagonist, WEB 2086 (2 mg/kg intravenously; Boehringer-Ingelheim, Ridgefield, CT).[45] The animals were separated into three groups. In each group, measurements of leukocyte adherence were performed at baseline, after 1 hour of ischemia, and after 1 hour of reperfusion. Animals in group 1 received no treatment and served as controls. Animals in group 2 received WEB 2086 before the induction of ischemia, and animals in group 3 received WEB 2086 10 minutes before reperfusion. There were no significant differences in baseline number of adherent leukocytes among the three groups of animals. Because complete flow arrest was obtained in the ischemic area, no change in the number of adherent leukocytes relative to baseline occurred at the end of 1 hour of ischemia in any group. In group 1 animals, the number of adhering leukocytes increased from 3.3 ± 0.7 (x = mean ± SEM) at baseline to 9.5 ± 1.0 at the end of 1 hour of reperfusion. Group 2 animals showed no significant difference between the preischemic baseline and values at 1 hour of reperfusion (3.5 ± 0.5 vs 3.8 ± 0.5). Group 3 animals exhibited no significant difference between the baseline values and those at one hour of reperfusion (2.3 ± 0.3 vs 2.3 ± 0.2). Similarly, in groups 2 and 3 there was no significant difference between the values measured in the ischemic areas and their time-matched nonischemic areas at the end of one hour of reperfusion (group 2: 3.8 ± 0.5 vs 3.0 ± 0.4; group 3: 2.3 ± 0.2 vs 3.2 ± 0.4). Interestingly, the group 1 animals exhibited a higher mean number of adherent leukocytes at the end of 1 hour of reperfusion compared with the preischemic baseline value (5.0 ± 1.8 vs 2.3 ± 0.4). This incremental trend was abolished by the administration of WEB 2086.

In a related set of experiments using the rat cremaster muscle preparation, significant leukocyte adhesion to postcapillary venular endothelium, as assessed by intravital microscopy, was noted after two hours of ischemia and 30 minutes of reperfusion in skeletal muscle.[46] In a separate set of experiments, the monoclonal antibody IB6 (Repligen Corporation, Cambridge, MA), which binds to the CD11B portion of the CD11/CD18 complex on neutrophils, was administered at reperfusion (2 mg/kg IV). Following 4 hours of ischemia, a significant decrease in leukocyte adhesion was noted after 30 minutes of reperfusion in the antibody treated group. This translated into a significant reduction in skeletal muscle parenchymal damage, as assessed by nitroblue-tetrazolium staining.

These data, as well as that from another laboratory,[44] support the hypothesis that PAF modulates leukocyte adhesion to endothelium in I/R and demonstrate that blockade of PAF receptors either before ischemia or before reperfusion reduces adhesion of leukocytes to endothelium. The efficacy of WEB 2086 in blocking the increased mean number of adherent leukocytes in the control nonischemic area also suggests that PAF may be released by the ischemic tissue (skeletal muscle) and act as a paracrine mediator to stimulate leukocyte adherence in normally perfused tissues. These data support the concept of using PAF receptor blockade in postischemic tissues as a therapeutic approach to limit I/R injury. In addition, our data derived from experiments using the IB6 antibody, as well as experiments using dissimilar methodologies reported by other investigators,[47,48,49] indicate that leukocytes play an important role in mediating parenchymal injury and microvascular dysfunction upon reperfusion of postischemic skeletal muscle. As the migration of leukocytes into the extravascular compartment, where their injurious effects upon the parenchyma are exerted, must be preceded by rolling and ultimately adhesion to endothelial cells,[50] therapeutic intervention directed against the CD11/CD18 leukocyte adhesion complex at the time of reperfusion provides an additional strategy to limit the severity of I/R injury in skeletal muscle, as well as in other tissues.

Acknowledgments

Original work from the authors' laboratory was supported by NIH grants HL 43146 and HL 25032, and grants-in-aid from the American Heart Association—New Jersey Affiliate.

References

1. Braquet P, Touqui L, Shen TY et al. Perspectives in platelet activating factor research. Pharmacol Rev 1987;39:97-145.
2. Archer CB, Page CP, Morley J et al. Accumulation of inflammatory cells in response to intracutaneous platelet activating factor (PAF-acether) in man. British J Dermatol 1985;112:285-290.
3. Pinckard RN. Platelet activating factor. Hosp Pract 1983;18:67-76.
4. Clarke PO, Hanahan DJ, Pinckard RN. Physical and chemical properties of platelet activating factor obtained from human neutrophils and monocytes and rabbit neutrophils and basophils. Biochem Biophys Acta 1980;628:69-75.
5. Lotner GZ, Lynch JM, Betz SJ et al. Human neutrophil derived platelet activating factor. J Immunol 1980;124:676-684.
6. Camussi G, Aglietta M, Coda R et al. Release of platelet activating factor and histamine. II. The cellular origins of PAF: monocytes, polymorphonuclear neutrophils and basophils. Immunol 1981;42:191-199.
7. Lewis RA, Goetzl EJ, Wasserman S et al. The release of four mediators of immediate hypersensitivity from human leukemia basophils. J Immunol 1975;114:87-92.
8. Arnoux B, Duval D, Beneveniste J. Release of platelet activating factor (PAF-acether) from alveolar macrophages by the calcium ionophore A23187 and phagocytosis. Eur J Clin Invest 1980;10:437-441.
9. Camussi G, Aglietta M, Malavasi F et al. The release of platelet activating factor from human endothelial cells in culture. J Immunol 1983;131:2397-2403.
10. Zimmerman G, McIntyre T, Prescott S. Production of platelet activating factor by human vascular endothelial cells: Evidence for a requirement for specific agonists and modulation by prostacyclin. Circulation 1985;72:718-727.
11. Chignard M, LeCouedic JP, Vargaftig BB et al. Platelet activating factor (paf-acether) secretion from platelets: Effect of aggregating agents. Br J Haematol 1980; 46: 455-464.
12. Barbaro JF, Zvaifler NJ. Antigen induced histamine release from platelets and rabbits producing homologous PCA antibody. Proc Soc Exp Biol Med 1966;122:1245-1247.
13. Schoenbechler MJ, Sadun EH. In vitro histamine release from blood cellular elements of rabbits infected with Schistosoma mansoni. Proc Soc Exp Biol Med 1968; 127:601-606.
14. Benveniste J, Henson PM, Cochrane CG. Leukocyte dependent histamine release from rabbit platelets. J Exp Med 1972; 136: 1356-1377.
15. Benveniste J. Platelet activating factor, a new mediator of anaphylaxis and immune comples deposition from rabbit and human basophils. Nature 1974;249:581-582.
16. Benveniste J, LeCoudic JP, Polansky J et al. Structural analysis of purified platelet activating factor by lipases. Nature 1977; 269:170-171.
17. Benveniste J, Tence M, Varenne P et al. Semi-synthesis and proposed structure of platelet activating factor (PAF): PAF-acether an alkyl ether analog of lysophospha-

tidylcholine. C R Seances Acad Sci D 1979;289:1037-1040.
18. Demopoulos CA, Pinckard RN, Hanahan DJ. Platelet activating factor. Evidence for 1-0-alkyl-2-acetyl-sn-glyceryl-3-phosphorylcholine as the active component. (a new class of lipid mediators) J Biol Chem 1979; 254:9355-9358.
19. Godfroid JJ, Heyman F, Michel E et al. Platelet activating factor (PAF-acether): Total synthesis of 1-0-octadecyl-1-0-acetyl sn glycero-3-phosphorylcholine. FEBS lett 1980;116:161-164.
20. Shen TY, Hwang SB, Doebber TW et al. The chemical and biological properties of PAF agonists, antagonists, and biosynthetic inhibitors. In: F. Snyder (ed) Platelet Activating Factors and Related Lipid Mediators. New York: Plenum Press; 1987:153-190.
21. Shukla SD. Inositol phospholipid turnover in PAF transmembrane signalling. Lipids 1991;26:1028-1033.
22. Hokin LE. Receptors and phosphoinositide generated second messengers. Ann Rev Biochem 1985;54:205-235.
23. Hwang SB, Lam MH, Dong SS. Ionic and GTP regulation of binding of platelet activating factor to receptors and platelet activating factor induced activation of GTPase in rabbit platelet membranes. J Biol Chem 1986;261:532-537.
24. Barzaghi G, Sarau HM, Mong S. Platelet activating factor induced phosphoinositide metabolism in differentiated U-937 cells in culture. J Pharmacol Exp Ther 1989; 248:559-566.
25. Honda ZI, Nakamura M, Miki I et al. Cloning by functional expression of platelet activating factor receptor from guinea pig lung. Nature 1991;349:342-346.
26. Prescott SM, Zimmerman GA, McIntyre TM. Platelet activating factor. J Biol Chem 1990;265:17381-17384.
27. Henson PM. Extracellular and intracellular activities of PAF. In: Snyder F (ed). Platelet Acivating Factors and Related Lipid Mediators. New York: Plenum Press; 1987: 255-267.
28. Shukla SD. Platelet activating factor receptor and signal transduction mechanisms. FASEB J 1992;6:2296-2301.
29. Hwang SB. Specific receptors of platelet activating factor, receptor heterogeneity, and signal transduction mechanisms. J Lipid Mediat 1990;2:123-158.
30. Linder ME, Gilma AG. G Proteins. Sci Am 1992;267:56-65.
31. Abdel-Latif AA. Calcium-mobilizing receptors, polyphosphoinositides and the generation of second messengers. Pharmacol Rev 1986;38:227-272.
32. Morrison WJ, Dhar A, Shukla SD. Staurosporine potentiates platelet activating factor stimulated phospholipase C activity in rabbit platelets but does not block desensitization by platelet activating factor. Life Sci 1989;45:333-339.
33. Braquet P, Hosford D, Braquet M et al. Role of cytokines and platelet activating factor in microvascular immune injury. Int Arch Allergy Appl Immunol 1989; 888: 88-100.
34. Koltai, Maytas, Hosford D et al. Role of PAF and cytokines in microvascular tissue injury. J Lab Clin Med 1992;119:461-466.
35. Durán WN, Dillon PK. Acute microcirculatory effects of platelet activating factor. J Lipid Mediat 1990;2:S215-S227.
36. Tomeo AC, Durán WN. Resistance and exchange microvessels are modulated by different PAF receptors. Am J Physiol 1991;261:H1648-H1652.
37. Tomeo AC, Egan RW, Durán WN. Priming interactions between platelet activating factor and histamine in the in vivo microcirculation. FASEB J 1991;5:2850-2855.
38. Butcher EC. Leukocyte endothelial cell recognition: Three (or more) steps to specificity and diversity. Cell 1991;67:1033-1036.
39. Springer TA. Adhesion receptors of the immune system. Nature 1990; 346: 3425-434.
40. Dillon PK, FitzPatrick MF, Ritter AB et al. Effect of platelet activating factor on leukocyte adhesion to microvascular endothelium. Time course and dose-response relationships. Inflammation 1988; 12:563-573.
41. Dillon PK, Ritter AB, Durán WN. Vasoconstrictor effects of platelet activating factor in the hamster cheek pouch microcirculation: Dose related relations and pathways of action. Cir Res 1988;62:722-731.

42. Dillon PK, Durán WN. Effect of platelet activating factor on microvascular permselectivity: Dose response relations and pathways of action in the hamster cheek pouch microcirculation. Cir Res 1988;62:732-740.
43. Yasuhara H, Hobson RW, Dillon PK et al. A new model for studying ischemia-reperfusion injury in hamster cheek pouch. Am J Physiol 1991;261:H1626-H1629.
44. Kubes P, Ibbotson G, Russel J et al. Role of platelet activating factor in ischemia/reperfusion-induced leukocyte adherence. Am J Physiol 1990;259:G300-G305.
45. Durán WN, Milazzo V, Hobson II RW. Microcirculatory dysfunction in ischemia/reperfusion injury: Role of platelet activating factor. J Vasc Surg 1993;18:113-114.
46. Ferrante RJ, Silva Jr MB, Resnikoff M et al. Inhibition of leukocyte adhesion at reperfusion decreases tissue damage in postischemic skeletal muscle. Surg Forum 1993;44:367-369.
47. Korthuis RJ, Grisham MB, Granger DN. Leukocyte depletion attenuates vascular injury in postischemic skeletal muscle. Am J Physiol 1988;254:H823-827.
48. Yokota J, Minei JP, Fantini GA et al. Role of leukocytes in reperfusion injury of skeletal muscle after partial ischemia. Am J Physiol 1989;257:H1068-1075.
49. Carden DL, Smith JK, Korthuis RJ. Neutrophil mediated microvascular dysfunction in postischemic canine skeletal muscle: role of granulocyte adherence. Circ Res 1990; 66:1436-1444.
50. Weiss SJ. Tissue destruction by neutrophils. N Engl J Med 1989;320:365-376.
51. Lapetina EG. Platelet-activating factor stimulates the phosphatidylinositol cycle. J Biol Chem 1982; 257:7314-7317.

CHAPTER 5

Complement Activation

Thomas F. Lindsay
Alexander D. Romaschin
Barry B. Rubin
Paul M. Walker

Ischemic injury when followed by reperfusion is associated with inflammation.[1] Initiation of inflammation in skeletal muscle or any reperfused tissue requires a stimulus which triggers the inflammatory response and recruits inflammatory cells, such as neutrophils, to the area of injury. This requires the production of chemoattractants generated in the tissues themselves such as arachidonic acid metabolites (leukotrienes, thromboxanes and hydroxyeicosatrienoic acids), modified plasma lipids, endothelial cell factors, cytokines and complement anaphylatoxins.[2-4] Several of these mediators have been shown to result in the initial rolling of neutrophils on endothelial cells, which is followed by firm adhesion.[5] The selectin family has been shown to play an important role in polymorphonuclear leukocyte (PMN) rolling,[6,7] while the CD11/CD18 (Mac-1) complex is critical for firm adhesion and diapedesis out of the postcapillary venule into the tissue.[8] Once within the extravascular compartment, neutrophils may adhere to parenchymal cells,[9] causing direct injury and/or cell lysis.[10] The complement system is unique as its activation attracts, activates and amplifies many aspects of the inflammatory reaction.

COMPLEMENT

The complement cascade is a phylogenetically conserved series of plasma proteins which are known to mediate several aspects of the inflammatory response.[3,11,12] The complement system consists of two pathways, the classical and alternate pathways. These pathways have different activating mechanisms; however they both result in the formation of the membrane attack complex leading to cellular lysis. The activation of either complement cascade will lead to production of the biologically active anaphylatoxins C3a and C5a. These are produced from the cleavage of native C3 and C5 by the

Ischemia-Reperfusion Injury of Skeletal Muscle, edited by Gary A. Fantini, M.D.;

respective C3 and C5 convertases, which are intermediate steps in the pathway of complement activation. These peptides mediate a variety of physiological effects including vasodilatation, increased vascular permeability, neutrophil chemotaxis, diapedesis, degranulation and oxygen-free radical production, as well as the translocation of the intracellular complement receptors to the neutrophil-plasma membrane. In addition, C3b and C4b serve as opsonins, thus facilitating phagocytosis of particles to which these molecules have attached. The complement cascade has four principle biological functions. The first biological function is that of cell lysis. The activated complement system can effect the killing of foreign organisms by osmotic cell lysis. The second major function is that complement proteins bind to the surfaces of foreign organisms or particles. These opsonins will bind to specific counterreceptors on macrophages and leukocytes. This facilitates phagocytosis and clearing of these organisms from the body. The third function of the complement system is augmentation, potentiation and promotion of the inflammatory response. The complement fragments will cause chemotaxis of inflammatory cells to the site of release. In addition to their chemotactic function, C3a and C5a stimulate neutrophil oxidative burst and mast cell degranulation, which potentiates the inflammatory response. The fourth function of the complement system is clearing of immune complexes from the circulation.

The complement cascade has several key properties which make it an important effector mechanism in the inflammatory response. Complement is a cascade system much like that of the coagulation cascade. Once activated, many of its components have enzymatic activity. One activated molecule at any step will result in a large amplification as each step in the pathway proceeds. While this is an important positive feedback loop, there are also multiple levels at which inhibitory proteins can interact in the cascade to prevent its inadvertent or improper activation. Another important property of the complement cascade is that it has both specific and nonspecific activators. Activation by antigen-antibody complexes, either several closely related IgG molecules or one IgM pentamere, can activate the classical pathway once the specific antibody and antigen have reacted. The alternate pathway is activated by direct binding of components to the surface of foreign organisms. In this way, the alternate pathway participates in nonspecific immunity of the host. A schematic representation of the complement cascade is depicted in Fig. 5.1.

THE CLASSICAL PATHWAY

The classical pathway of complement is initiated by the interaction of C1 with an antigen-antibody complex. C1 will not interact with either IgG or IgM unless they have bound their specific antigen. C1q is the subunit of C1 that binds to the immunoglobulin molecules. C1q binds either to the CH3 domain of IgM or the CH2 domains of IgG which reside in the Fc portion of the antibody. A single C1q molecule must bind to at least two Fc portions of an immunoglobulin. IgM is secreted as a pentamere containing five separate Fc regions, therefore a single IgM molecule bound to an antigen can trigger complement activation. Thus, IgM is an efficient activator of complement and is also a complement fixing antibody. C1q can also bind to adjacent IgG molecules in order to be activated. C1q is uniquely able to bind these different Fc regions of immunoglobulins because it consists of six identical subunits that each have projecting heads. The heads are available to interact with the CH components of the Fc portion of the immunoglobulins. Once C1q is bound to an activating immunoglobulin, a tetramere which consists of two C1r and two C1s molecules is added. After these are bound, a serum esterase-type enzyme is formed which acts on the next two components, C4 and C2. C4, like several of the complement proteins, has an internal thioester bond. Once C4b is generated, this reactive form of C4b is called metastable C4b. Many of these C4b molecules may react with water at their thioester site thereby inactivating the C4 complex. Alternatively, some of the C4b thioester bonds may esterify to cell surface proteins or carbohydrates stabilizing the C4 complex. In the presence of magnesium, C2 can bind to the C4b molecules that are bound to the cell surface proteins or carbohydrates. Once C2 is added, it is split by a C1s molecule generating C2b which may diffuse away. The C2a fragment remains complexed with C4b resulting in a C4b2a complex, which is the C3 convertase of the classical pathway (Fig. 5.1). This C3 convertase requires calcium and magnesium and results in the conversion of C3 into C3a and C3b. This convertase is able to bind C3 and results in its proteolytic cleavage into C3a and C3b. C3b has an internal thioester which may form a covalent bond with the cell surface or be inactivated by reacting with water. The addition of C3b to the preexisting C4b2a complex results in formation of the C5 convertase of the classical pathway. This C5 convertase will result in the cleavage of C5 and initiate the formation of the membrane attack complex. In addition, in vitro experiments have demonstrated that the classical pathway may be activated by infectious organisms such as Mycoplasma or molecules such as DNA, heparin and chondroitin sulfate.

THE ALTERNATE PATHWAY

The alternate pathway is in a constant low-level state of activation due to the spontaneous hydrolysis of the internal beta cyteinyl gamma glutamyl thioester bond in the D domain of the alpha chain of C3. Thioester decay results in the transient formation of the enzyme complex necessary for the activation of C3 convertase. This exposes the metastable binding site of nascent C3b and there are potential activators and

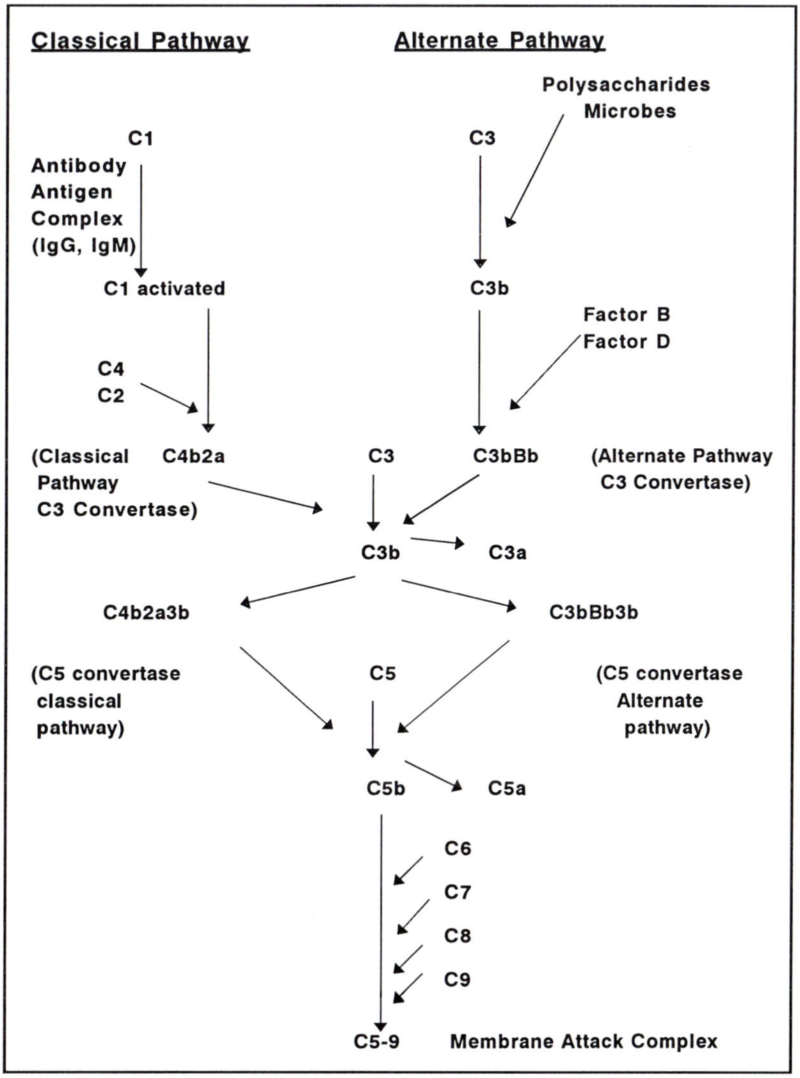

Fig. 5.1. Molecular organization of the complement cascades. Antibody antigen complexes will initiate C1q binding to activate the classical pathway. The alternate pathway may be activated spontaneously by the classical pathway or activating surfaces such as microbial cell walls. Each pathway generates distinct C_3 and C_5 which leads to the formation of the membrane attack complex.

inactivators of this pathway. The alternate pathway of complement activation also requires the production of C3 and C5 convertase in order to generate the membrane attack complex. The alternate pathway uses a different series of activators and different series of proteins than the classical pathway in order to produce the C3 and C5 convertases. The classical pathway may be the initiator of the alternate pathway, as the first C3b molecules often come from classical pathway activation. In the alternate pathway, C3b interacts with a second molecule of C3 which has undergone slow spontaneous hydrolysis (called the tickover of C3). This process is thought to be ongoing at low levels in the circulation at all times. C3b may be inactivated by hydrolysis if it is deposited on autologous cell surfaces which have endogenous inhibitors of complement activation incorporated within their cell surface membrane. Should C3b bind to the surface of a microbe which is unable to inactivate C3b, then factor B will become complexed to the molecule. Tissues which allow propagation of the alternate pathway appear to do so by causing a conformational change in bound C3 molecules, which discourages inhibitory inactivation and permits the binding of factor B. Next, factor D, which is thought to circulate in a catalytically active form, is added to the complex forming C3bBb. This complex is relatively unstable and will rapidly decay unless properdin is added to the complex. This forms C3b convertase which has a positive feedback effect, as the formation of C3 convertase will generate more C3b which may further trigger the pathway. Once additional C3b is added to the complex forming C3bBb3b, which is the C5 convertase. This C5 convertase will also initiate the hydrolysis of C5 resulting in the progressive formation of the membrane attack complex.

MEMBRANE ATTACK COMPLEX

Either of the C5 convertases will cleave C5 into its a and b fragments. The C5b component will allow binding of C6 and C7, the latter of which will insert the complex into the lipid bilayer of a plasma membrane. This molecule becomes a receptor for C8 and once C8 is added, lysis of the cell may begin. As many as 12 to 15 C9 molecules may then be added to the complex, with resulting full lytic capacity. This combination of proteins undergoes hydrophilic association within the plasma membrane and causes a weakening of the membrane structure and formation of transmembrane channels. This may cause cell death by osmotic lysis with a failure to maintain critical intra- and extracellular gradients. Scanning electron microscopy has suggested that the membrane attack complex appears to act as a pore within the cell membrane allowing the collapse of ionic and osmotic gradients. It is the collapse of these gradients that ultimately results in the demise of the cell. In addition, it should be noted that because of the amplification mechanisms involved in both pathways, activation of complement results in the production of many membrane attack complexes on the cell surface so that the combination of these pores leads to gradient collapse.

As outlined in Figure 5.1, there are several by-products of the complement cascade. Perhaps best known of these are the anaphylatoxins. C3a, C4a and C5a comprise this group of molecules and they have been so named because they effect a rapid increase vascular permeability. These molecules can interact with receptors on a variety of cells including basophils, lymphocytes, monocytes, macrophages, neutrophils and mast cells. Once bound to mast cells and basophils, granules are exocytosed releasing histamine and other mediators which result in smooth muscle contraction and increased vascular permeability. C5a is a powerful chemoattractant which activates neutrophil respiratory burst, resulting in granule discharge and an increase in their adhesive ability. A second component is the binding of both activated and inactivated complement fragments to the cell surface. Specifically, iC3b is a counter-receptor for the neutrophil CD11b/CD18 (Mac-1) receptor found on neutrophils and monocytes. This will enhance the binding of neutrophils to complement coated molecules, thereby assisting in their phagocytosis.

REGULATION OF THE COMPLEMENT CASCADE

The complement pathway contains an amplification mechanism which has several positive feedback steps, resulting in the formation of several proteolytic enzymes which can generate large numbers of activated complement fragments. This cascade requires tight regulation in order to prevent inadvertent activation and damage to the host. There are several soluble factors found in the circulation which inactivate various aspects of the complement cascade. These include the C1 inhibitor, the C4 binding protein, factor H and factor I. In addition, there are several membrane proteins which are found within the membrane of most of the cells of the myeloid line as well as platelets and mast cells. These include the complement type 1 receptor, membrane cofactor protein and the decay accelerating factor. Molecules which inhibit membrane attack complex formation include homologous restriction factor and the membrane inhibitor of reactive lysis. These two factors are thought to be key proteins in protecting normal bystander cells from adjacent areas where complement activation and cell lysis are ongoing. The complement receptor type 1 (CR1) has at least three important functions. First, it can inhibit C3 convertase activity, second it enhances the ability of leukocytes to ingest C3b or C4b coated particles and thirdly, it facilitates the clearance of immune complexes from the circulation. CR1 is present on erythrocytes and in smaller quantities on neutrophils, monocytes, macrophages, eosinophils and lymphocytes. CR1 is an important regulatory protein as it can inhibit both the classical and alternate pathways of complement activation.

In summary, the complement cascade can be extremely beneficial to the host as a mechanism of activating and potentiating the inflammatory response. Its by-products have the ability to attract and activate neutrophils and other inflammatory cells, to release histamine causing vasodilatation and increased vascular permeability, thereby allowing further complement proteins to diffuse out of the circulation to an area of inflammation. Failure to control this very powerful arm of the inflammatory system following activation can result in significant damage to the host tissues.

COMPLEMENT ACTIVATION FOLLOWING TISSUE ISCHEMIA

Along with the mechanisms outlined above, other methods of activating the complement pathway have been described. Antibody- independent activation of the classical pathway has been demonstrated in vitro, resulting from the binding of C1q to molecules of mitochondrial origin which are exposed after myocardial ischemia.[13] C1q binds to these subcellular constituents which have been identified in cardiac lymph after myocardial ischemia.[14] These studies in myocardial ischemia have prompted further investigations into the role of complement in mediating myocardial injury following ischemia/reperfusion (I/R). Complement has been difficult to specifically inhibit in animal species and for this reason many investigators have depleted complement by administration of cobra venom factor, which depletes complement proteins by activating them. The size of myocardial infarctions were studied in baboons randomized to either cobra venom factor, saline or ibuprofen.[15] Treatment of animals just prior to reperfusion with either cobra venom factor or ibuprofen resulted in significant reduction in the accumulation of neutrophils in the infarcted area, as well as histological evidence of a reduction in transmural infarction. There was no significant difference between the cobra venom factor- or ibuprofen-treated animals with respect to the reduction in infarct size. The reduction in infarct size was comparable with both treatments when compared to saline. However, in the ibuprofen and saline treated groups, there was extensive localization of C3, C4 and C5 in all infarct sites. In cobra venom-treated animals, only small amounts of C4 could be localized. The depletion of complement in this study reinforced the hypothesis that complement mediates reperfusion injury following myocardial ischemia and can attenuate the size of subsequent infarction.

Research into the role of complement in postischemic myocardial inflammation and necrosis was furthered when the complement type 1 receptor gene was isolated.[16] This endogenous protein, designated CR1 (CD35), is a powerful inhibitor of both the classical and alternate pathway activities. Its function in vivo is limited by its restriction to a few cell types such as erythrocytes and cells of the myeloid cell line. CR1 is a transmembrane protein and the gene has been altered to delete the intracellular and transmembrane components. This altered molecule, called soluble CR1 (sCR1), retained its ability to inhibit both the alternate and classical pathways of complement pathways of complement activation. It appeared that the alternate pathway C3 and C5 convertases were equally susceptible to the inhibitory affect of sCR1, while the classical pathway C5 convertase was more readily suppressed than the classical C3 convertase.

sCR1 was administered to rats immediately before left coronary artery suture ligation designed to produce a myocardial infarction.[16] After 35 minutes, sutures were released and the animals were allowed to survive for 7 days. At that time, infarct size and neutrophil sequestration within areas of infarction were quantitated. Administration of sCR1 significantly reduced the size of the myocardial infarctions from 16% to 9%, in addition to lowering the incidence of transmural infarction. When serial sections were examined and the number of neutrophils quantified, a dramatic reduction in the number of leukocytes sequestered in the myocardium of those animals treated with sCR1 was noted. This indicated that the suppression of complement activation was associated with decreased inflammatory cell accumulation in the heart, but did not appear to impair healing of the infarcted area. Immunoperoxidase staining for the membrane attack complex was noted in infarcted areas primarily along the endothelium of capillaries and venules, but not on the myocardial fibers. Those animals treated with sCR1 had little or no deposition of the C5b-9 complex detected in these regions. This study suggests that the activation of complement plays a key role in neutrophil infiltration and the development of myocardial necrosis following ischemia. In addition, it demonstrated that endothelium was the initial site of complement deposition. The production of C5a, which must precede C5b-9 complex deposition, would activate leukocytes causing up-regulation of their cellular receptors CR1 and CR3. The latter receptor, CR3, also known as the CD11b/CD18 (Mac-1) receptor complex, binds iC3b and thus would be important in promoting attachment of neutrophils to damaged tissues. These experiments suggest that complement plays an important role in reperfusion injury of previously ischemic myocardium.

COMPLEMENT ACTIVATION AND SKELETAL MUSCLE

Many fewer studies have examined the role of complement in I/R injury of skeletal muscle. Heideman and colleagues demonstrated that the infusion of a homogenized pellet of skeletal muscle tissue into dogs resulted in severe shock and activation of the complement pathway, as measured by a reduction in the CH50.[17] In a clinical study, concentrations of C3a and C5a were measured in patients who had ischemic limbs.[18] Increased levels of these anaphylatoxins were measured preoperatively and returned to normal following amputation. These studies indirectly suggested that complement was activated by ischemic skeletal muscle. It was suggested that the enhanced activity of these anaphylatoxins may be responsible for abnormalities of respiratory, hepatic and renal function in patients with extensive muscle ischemia. There were three patients who required amputation at a higher level for further ischemia of their limbs. In these three patients, the levels of C3a had initially fallen after their first operative intervention but they were elevated once again at the time of reoperation, and fell once again following the second operation. These findings suggested that complement is activated by acute limb ischemia; however, there was no direct evidence that the ischemic skeletal muscle was responsible for the elevation of complement anaphylatoxins levels. Neither was evidence provided that complement fragments were deposited in the ischemic skeletal muscle.

Neutrophils have been shown to accumulate in skeletal muscle in response to I/R, and to play an important role in mediating tissue injury.[19,20] Specifically, reperfusion with neutrophil-depleted blood has been shown to reduce the initial increase in microvascular permeability and resistance in postischemic skeletal muscle.[21] Indeed, it is apparent that neutrophils occupy a central position in the pathogenesis of I/R injury in skeletal muscle[22-25] and cardiac muscle.[26,27] The infusion of zymosan (an activator of complement and neutrophils) activated plasma into an in vitro preparation of canine skeletal muscle via a perfusion bath has been shown to result in significant increases in vascular resistance and muscle neutrophil sequestration and decreases in solvent drag coefficient.[28] This injury has been successfully attenuated by antineutrophil serum or a monoclonal antibody against the neutrophil CD18 receptor.[28] These results suggested that zymosan activated plasma induced neutrophils to adhere to skeletal muscle vascular endothelium, resulting in microvascular dysfunction. These data suggest that if complement is activated by ischemic skeletal muscle, it can induce many microvascular changes via modulation of neutrophil activity.

ACTIVATION OF THE ALTERNATE COMPLEMENT PATHWAY FOLLOWING EXPERIMENTAL MUSCLE ISCHEMIA

A series of experiments was undertaken to determine if complement was activated and if neutrophil sequestration developed following an in vivo preparation of canine skeletal muscle ischemia.[29] The experimental protocol consisted of isolating the canine gracilis muscle on a single vascular pedicle. The muscles were then subjected to 5 hours of normothermic ischemia followed by 48 hours of reperfusion. Neutrophil sequestration was evaluated by two methods. First, in control and I/R animals, the gracilis muscle arterial inflow and venous effluent blood was collected in EDTA prior to ischemia and during perfusion. Blood flow at each sampling point was determined by timed venous collections and the number of neutrophils in each sample was measured using an automated analyzer. Wright's stain was also employed on random samples of the blood to quantify the number of PMNs in the sample. Neutrophil sequestration in the muscle was calculated by subtracting the venous neutrophil count from that of the arterial and multiplying by the blood flow and dividing by the preischemic muscle weight. Thus, the number of neutrophils being sequestered in the muscle was calculated by integrating the area under the curve of neutrophil sequestration over time. The second method of quantifying skeletal muscle neutrophil sequestration was performed by biopsying the muscles at various intervals and analyzing the tissue for myeloperoxidase (MPO) activity.

To determine if the alternate or classical pathways of complement were activated, assays were performed to quantify the levels of factor B and C4. Factor B is a critical component of alternate pathway activation (Fig. 5.1). Secondly, C4 is only consumed by classical pathway activation. Thus, changes in the relative levels of these metabolites would be indicative of significant consumption and imply activation of one or both pathways during experimental muscle ischemia. Analysis of C4 and factor B were performed on arterial blood taken at the end of ischemia and during the period of reperfusion. Animals which did not undergo ischemia served as controls.

The quantitation of factor B was based on the ability of rabbit red blood cells (RBC) to activate the canine alternate pathway cascade, leading to membrane attack complex formation and RBC lysis. The quantitation of C4 was based on the ability of canine C4 to reconstitute C4 deficient serum to lyse antibody-coated sheep RBC. The activity of factor B or C4 in the arterial blood drawn just before the end of ischemia was defined as one hemolytic unit in each animal. This was related to the volume of plasma required to cause a 63% hemolysis of rabbit or antibody-coated sheep RBC, which is a standard method of quantitating complement hemolytic activity. To determine if systemic factor B or C4 levels changed during reperfusion, the hemolytic activity of the arterial plasma drawn at the end of the ischemic interval and during reperfusion was compared. The relative factor B and relative C4 activity of samples collected as reperfusion proceeded was determined by the ratio of the volume of plasma required for 63% hemolysis in ischemic versus reperfusion samples. This was corrected for dilutional changes using the total plasma protein concentration. The data for the factor B and C4 levels was analyzed using a repeated measures analysis of variance (RM-ANOVA) followed by unpaired t testing of samples. There were 25 muscle specimens which underwent testing; 19 were exposed to 5 hours of ischemia and 6 served as controls. Muscle necrosis in the I/R group averaged 45% ± 6.1%, as compared with 0% ± 0.0% in the control group ($P < 0.0003$).

Neutrophil Sequestration in Skeletal Muscle

Muscles that were exposed to I/R had a significantly higher sequestration of neutrophils than controls during the initial 4 hours of reperfusion (Fig. 5.2). A 4- to 5-fold increase in neutrophil sequestration in the ischemic muscles was observed compared to time-matched controls over the first 3 hours of reperfusion. The difference was not statistically significant in the fourth hour of reperfusion and no relationship between neutrophil sequestration and muscle necrosis in the I/R group was noted. The results of MPO analysis performed on surface muscle biopsies are displayed in Figure 5.3.

A significant rise in MPO activity was noted after 48 hours of reperfusion. In muscles subjected to I/R, MPO activity obtained prior to and at the end of ischemia were unchanged. During the first 4 hours of reperfusion, no significant increase in muscle MPO noted. At 48 hours, MPO was elevated 97-fold in the I/R group compared to 11-fold in the control muscles. At earlier time points, increases in MPO activity occurred in both I/R and control groups. Although the increase was consistently greater in the I/R group than in the controls, this did not achieve statistical significance at

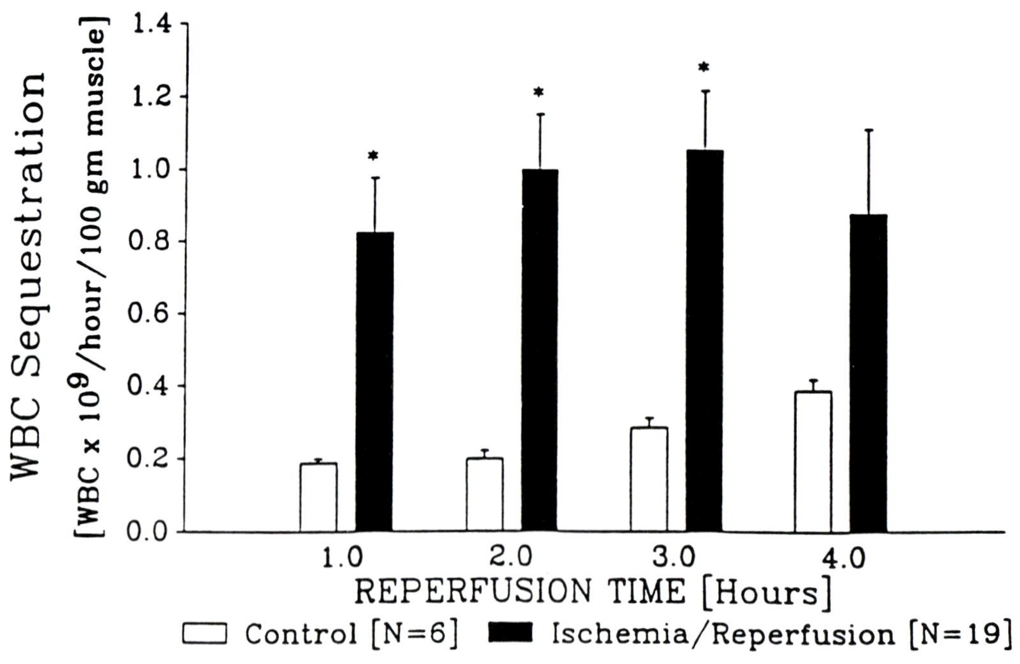

*Fig. 5.2. Neutrophil sequestration during the first, second, third and fourth hours of reperfusion in vivo after five hours of global ischemia. $p = 0.0021$ by RM-ANOVA; *$p < 0.05$ versus time matched, nonischemic controls by unpaired t test. Reprinted with permission from Rubin,* Complement Activation of White Cell Sequestration in Postischemic Skeletal Muscle. *Amer J Physiol 1990; 2(59):H525-531.*

the early time points. Once again, no relationship between muscle necrosis and MPO was noted at any time point.

Factor B Analysis

Repeated measures analysis of variance was applied to the differences in relative factor B activity (by defintion equal to 1.0) between samples taken at the end of ischemia and during reperfusion for each animal. In the control group, there was no change in relative factor B activity during the first four hours of reperfusion (Fig. 5.4). In the I/R group after 15 minutes of reperfusion, the relative factor B activity had decreased from 1.0 to 0.82 ± 0.04 ($P < 0.05$). No evidence of any further changes in factor B levels between 15 and 240 minutes of reperfusion were noted. These results indicate that postischemic skeletal muscle comprising 0.5% of total body weight was able to stimulate a consumption of 18% of the systemic factor B pool during the first 15 minutes of reperfusion. After 48 hours of reperfusion, the relative factor B activity in the I/R group had increased by 62%. In control animals, factor B had also increased by approximately 27%. This suggested that the surgical procedures stimulated an increase in systemic factor B production.

C4 Analysis

There was no overall change in relative C4 activity during the first 4 hours of reperfusion ($P = 0.19$ by RM-ANOVA). In seven of the nine animals studied, the relative C4 activity did not deviate more than 10% from the preischemic values. Two animals, however, showed relative decreases in C4 activity to 0.54 and 0.77 after 15 minutes and remained depressed for the first four hours of reperfusion. The levels of factor B after 48 hours were unchanged compared to preischemic values. There was no change in C4 activity in the control group animals. Thus, unlike the systemic increase in factor B levels, the surgical procedure did not appear to result in increased C4 production.

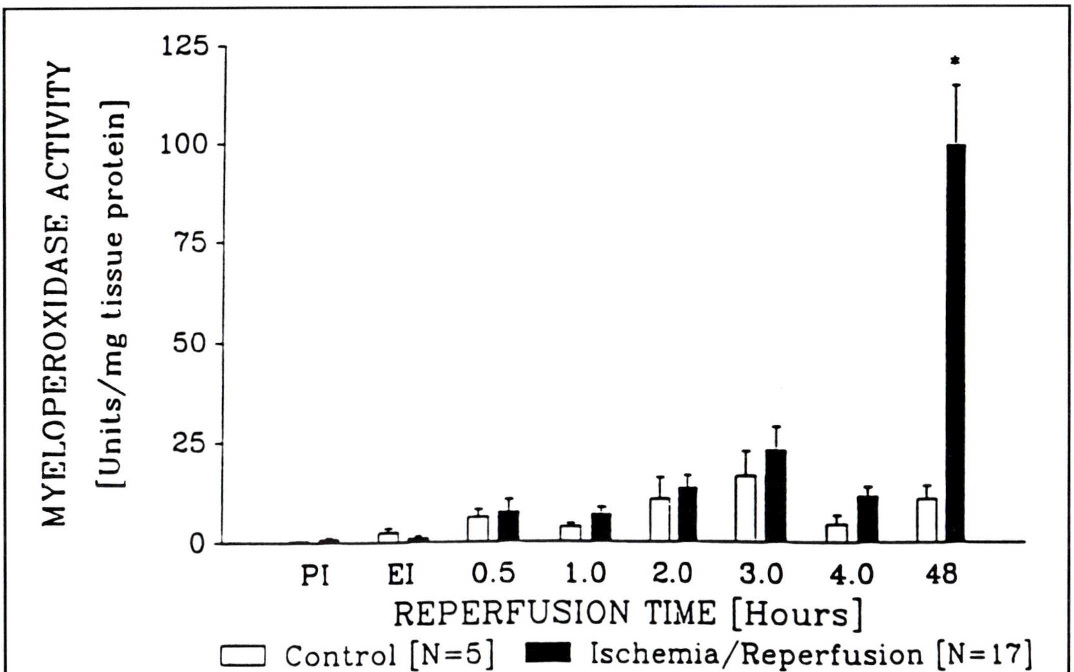

*Fig. 5.3. Myeloperoxidase activity in surface tissue biopsies during the initial 48 hours of reperfusion in vivo. PI, preischemia; EI, end ischemia; p = 0.0004 by RM-ANOVA; *p = 0.00001 versus time matched, nonischemic control by unpaired t test. Reprinted with permission from Rubin,* Complement Activation of White Cell Sequestration in Postischemic Skeletal Muscle. *Amer J Physiol 1990; 2(59):H525-531.*

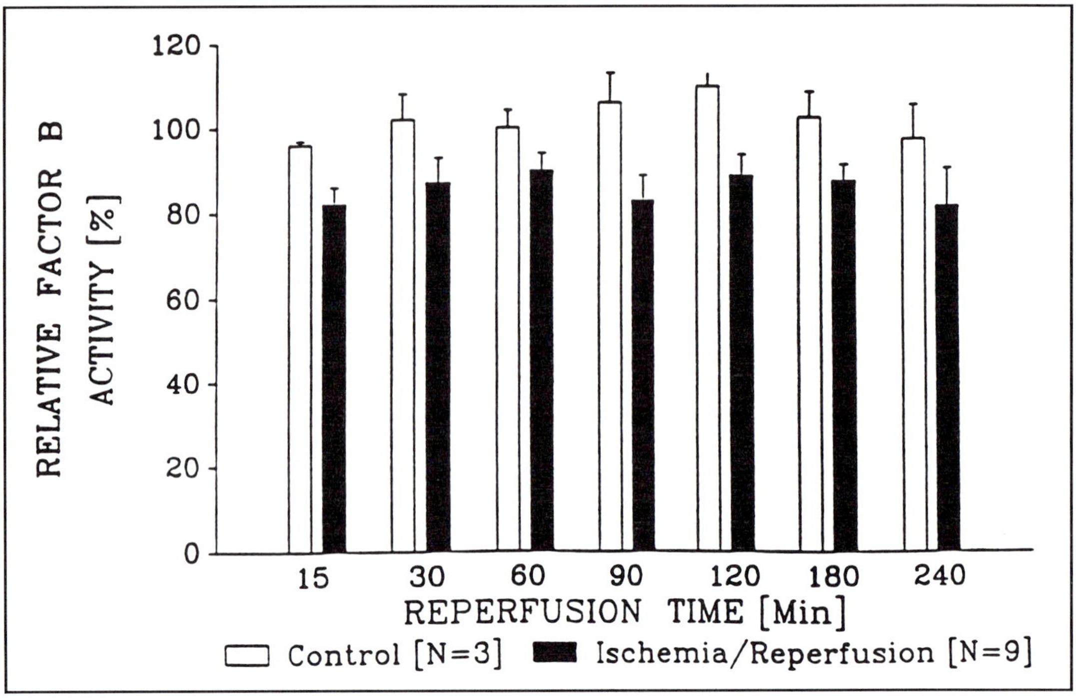

Fig. 5.4. Systemic factor B activity during reperfusion. The decline in factor B activity with time (p = 0.023 by RM-ANOVA) demonstrates that ischemia/reperfusion potentiated the activation of the alternative complement cascade. Reprinted with permission from Rubin, Complement Activation of White Cell Sequestration in Postischemic Skeletal Muscle. *Amer J Physiol 1990; 2(59):H525-531.*

This study used two different methods to quantify neutrophil sequestration in canine skeletal muscle following I/R. The results from the arterial-venous measurements indicated that neutrophils are rapidly and extensively sequestered in skeletal muscle during the initial few hours of reperfusion following five hours of ischemia. Tissue MPO activity increased 7-fold after 1 hour and 13-fold after two hours of reperfusion in the I/R group. However, these differences were not statistically significant, as there were similar increases in MPO activity in the control groups. Thus, the observed increases in MPO activity may have been a result of surgical manipulation which resulted in sequestration of these neutrophils. These biopsies were taken on the surface of skeletal muscle. Previous investigations had documented that skeletal muscle necrosis begins in the center of the muscle and proceeds toward the outside as the duration of ischemia increases. Similarly, neutrophil sequestration may begin toward the center of the muscle and progress outward in a wave as well. Thus, it is possible that the biopsies taken at the surface may represent a sampling error, and hence did not reveal the degree of neutrophil sequestration indicated by the arterial-venous differences.

At the end of the 48 hour period of reperfusion, a 97-fold increase in tissue MPO activity was noted compared to preischemic levels. This suggests that there is ongoing sequestration of neutrophils in the postischemic muscle long after the initial period of reperfusion. This may explain why interventions which only treat skeletal muscle or other organs during the first hour or two of reperfusion may fail after a prolonged period of reperfusion. The second important result of this study is that the alternate pathway of complement is involved in the inflammatory response to I/R injury in skeletal muscle. The reduction in factor B that was observed suggests that the complement cascade has resulted in the formation of C3 and C5 convertase enzyme complexes. These complexes will act upon both C3 and C5 leading to the biologically active peptides C3a and C5a. The production of these peptides has been shown to be important in the chemotaxis of neutrophils, upregulation of CD18 receptor adherence to the endothelium, and production of oxygen-free radicals with neutrophil degranulation. The combination of hydrogen peroxide production via NADPH oxidase, and the release of MPO will result in the production of hypochlorous acid as well as other oxidants, and the combination of these oxidants and neutrophil proteases are thought to result in tissue injury.[10] Thus, the rapid consumption of factor B noted in this study demonstrates that complement is activated quickly after skeletal muscle reperfusion and may be one of the initiators of neutrophil recruitment into the muscle resulting in progressive muscle swelling and necrosis. The significant elevation noted in factor B levels at 48 hours following muscle I/R suggest that synthesis of factor B, but not C4, is part of the acute phase systemic response initiated by skeletal muscle ischemia.

COMPLEMENT BLOCKADE REDUCES SKELETAL MUSCLE PERMEABILITY AFTER ISCHEMIA

The ability to study the role of complement directly following skeletal muscle ischemia has been hampered by the lack of specific inhibitors. As soluble CR1 (sCR1) had been reported to attenuate myocardial I/R injury,[16] the effect of sCR1 was evaluated using a model of bilateral hind limb tourniquet ischemia.[30] Male rats were exposed to 4 hours of tourniquet ischemia of the lower limbs followed by four hours of reperfusion. Just prior to reperfusion, animals were injected with I^{125} labelled bovine serum albumin (BSA), which was used to quantify changes in both skeletal muscle and lung permeability. Tissue from the upper and lower limbs was subjected to drying at 70°C for 72 hours, in order to determine alterations in the wet to dry weight ratios. The relative muscle permeability index was estimated by dividing the I^{125} BSA counts per minute per gram dry weight of muscle tissue by the I^{125} BSA counts per minute per gram of blood. Lung and muscle sections were assayed for MPO activity using a standard assay. Complement activity was quantified by lytic assays for both the classical and alternate pathways. Lung permeability was assessed by repeated lavage of one lung and measurement of the amount of I^{125} in the lavage fluid. The time course of the alterations in muscle permeability index was initially determined (Fig. 5.5). Following 4 hours of tourniquet ischemia and 30 minutes of reperfusion, a significant increase in muscle permeability was noted in the ischemic hind limb muscle compared to sham hind limb muscle or upper limb muscle in sham or ischemia animals. The muscles of the upper limb served as a control both in sham and ischemic animals to ensure that there was no systemic change in the muscle permeability in the ischemic animals. The results demonstrated that only the muscles subjected to ischemia were injured in this experimental protocol. Figure 5.5 demonstrates the progressive changes in muscle permeability index over 4 hours of reperfusion. There was a significant rise in permeability to I^{125} BSA by 30 minutes and this rise continued 1 to 2 hours. At 2 hours, there was a marked increase in the slope of the curve of the permeability index which continued over the last 2 hours of reperfusion. No changes were noted in the muscle permeability index of any of the nonischemic muscles (Fig. 5.5). These data suggest that a large increase in the transudation of fluid across the capillary membrane occurs very early following reperfusion. The alterations in the wet to dry weight that occurred following hindlimb tourniquet ischemia at 30 minutes of reperfusion revealed a significant increase in muscle wet to dry weight ratio and it continued to increase during the remaining $3^1/_2$ hours of reperfusion (Fig. 5.6). At 4 hours of reperfusion the wet to dry weight ratio was significantly higher compared to the 30 minute time point. While change in permeability to fluid occurs very early, there are progressive changes in the permeability to protein as documented by the changes in I^{125} permeability index. These two events follow radically different time

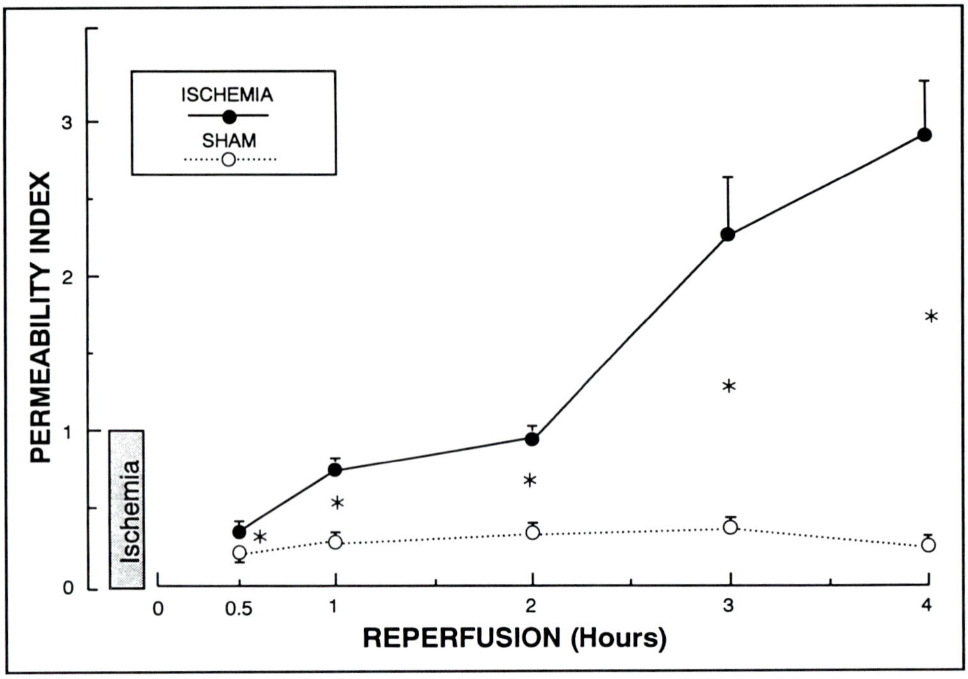

*Fig. 5.5. Changes in skeletal muscle permeability index with increasing duration of reperfusion in sham and ischemic animals. Significant elevations in ischemic hind limb muscle were observed compared with sham animals and upper limb control muscles. *$p < 0.05$ versus sham at the same time point. Reprinted with permission from Lindsay, TF.* The Albumen Leak after Lower Torso Ischemia Reperfusion. Ann Surg 1992; 216.

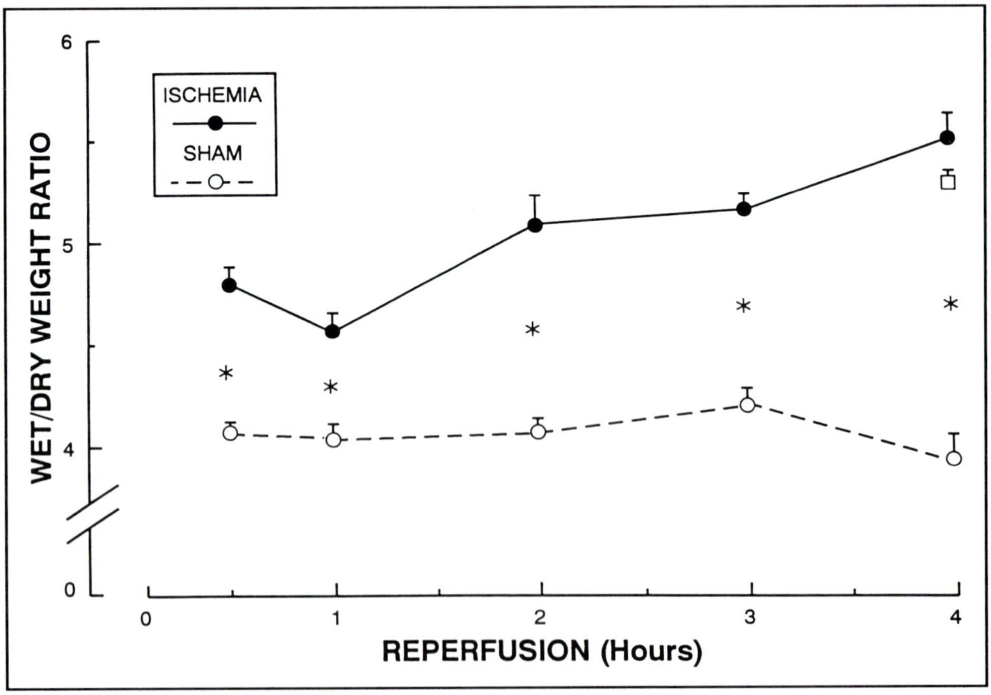

*Fig. 5.6. Hind limb muscle wet to dry weight ratio remained unchanged in sham but rose in ischemic/reperfused animals as the duration of reperfusion increased. *$p < 0.05$ compared with sham at the same time point. The wet to dry weight ratio for the ischemic muscle treated with 6 mg sCR-1 is shown as an open box, and was unaffected by sCR-1 treatment when compared with untreated ischemic hind limb muscle. Reprinted with permission from Lindsay, TF.* The Albumen Leak after Lower Torso Ischemia Reperfusion. Ann Surg 1992; 216.

courses of injury, suggesting that each of these events is mediated by separate and distinct inflammatory mechanisms and mediators in postischemic skeletal muscle.

The systemic effects of skeletal muscle I/R injury were examined by measuring lung permeability index and lung MPO content. The results demonstrated that there was roughly a doubling of lung MPO activity in animals subjected to I/R compared to sham and also a doubling of the lung permeability index. In contrast, no alterations in muscle MPO activity were noted during either ischemia or reperfusion.

The lytic assays that were performed demonstrated a significant drop in the lytic activity of the classical pathway at end-ischemia (Fig. 5.7). There was also a reduction in alternate pathway activity, however, this did not reach statistical significance. It is possible that a low level of blood flow via the medullary artery persisted despite tourniquet ischemia. The tourniquet cannot compress the medullary artery in the head of the femur and this small amount of blood flow may have resulted in the reduction in lytic activity noted at the end of the ischemic period. In animals that received phosphate buffered saline as a treatment, there was a continued decline in the classical pathway lytic activity during reperfusion while the alternate pathway lytic activity was unchanged.

In a separate group of animals, sCR1 was administered by a bolus injection three hours and 55 minutes after the onset of ischemia. At four hours the tourniquets were released and the animals were sacrificed four hours later. In animals treated with 1 mg of sCR1, there was a significant reduction in muscle permeability index compared to those untreated muscles (Fig. 5.8). When higher doses of sCR1 were administered, 3 and 6 mg per animal, a step-wise reduction in the muscle permeability index was noted. At the 6 mg dose, there was a greater

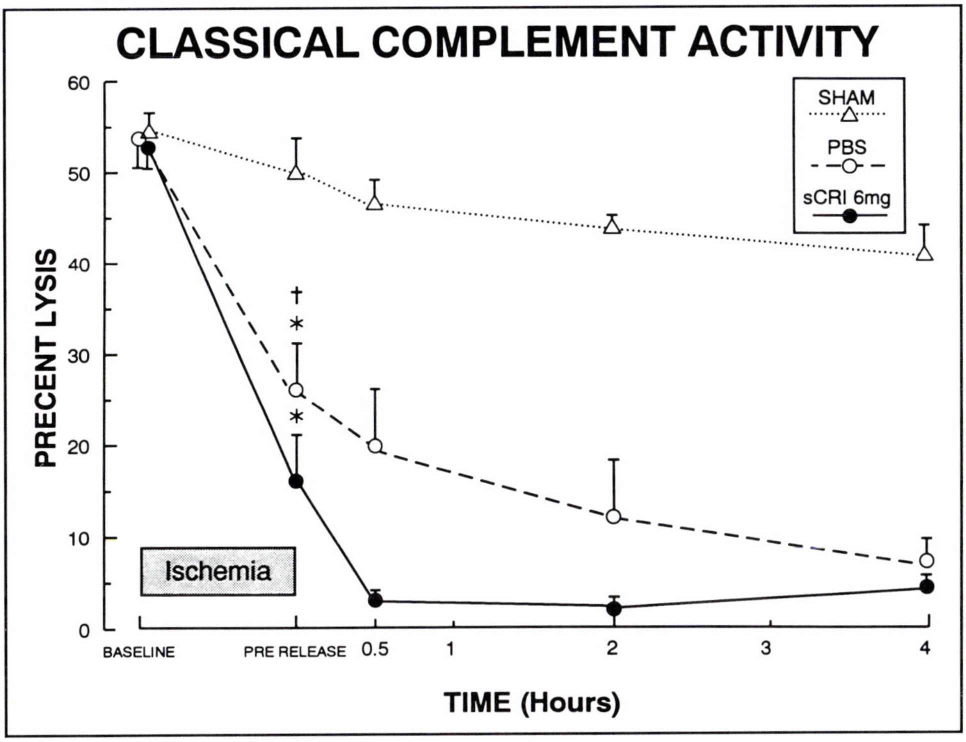

*Fig. 5.7. Rat classical complement pathway lytic capacity decreased significantly after bilateral hind limb ischemia. After the prerelease sample was taken, sCR-1 was administered and reperfusion initiated. PBS, phosphate-buffered saline; *p < 0.05 versus sham at the same time point, †p < 0.05 versus baseline sample in the respective group. Reprinted with permission from Lindsay, TF.* The Albumen Leak after Lower Torso Ischemia Reperfusion. Ann Surg 1992; 216.

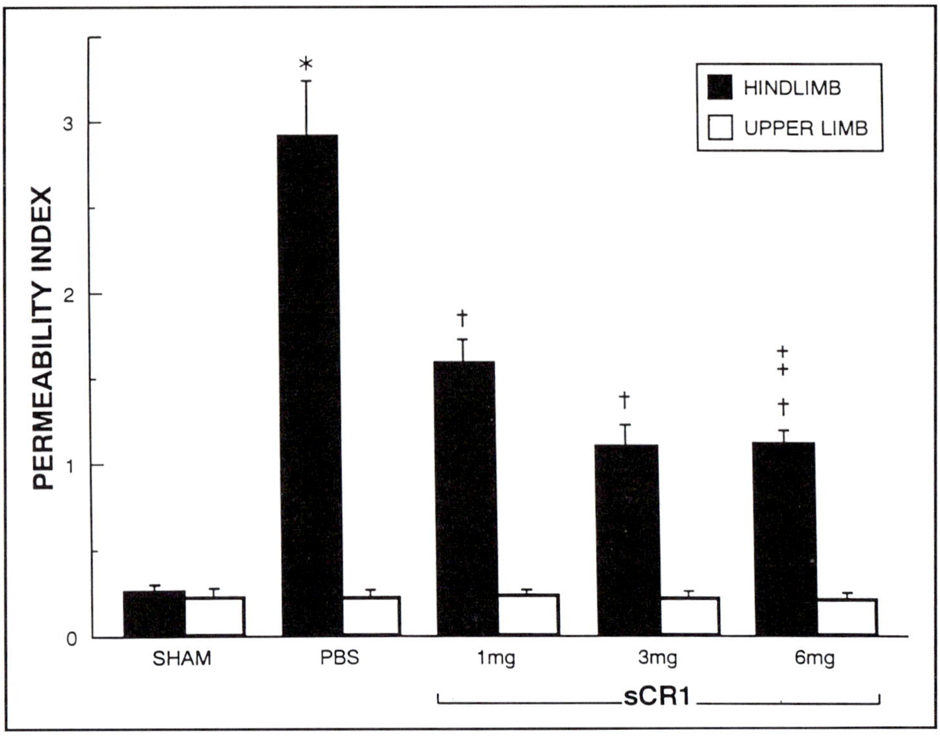

*Fig. 5.8. Treatment with sCR-1 reduced the permeability index in hind limb muscle after 4 hours of bilateral hind limb ischemia and 4 hours of reperfusion. The 6 mg dose was significantly better than the 1 mg dose. PBS, phosphate-buffered saline; *p < 0.001 versus sham, †p < 0.01 versus PBS, ‡p < 0.05 versus 1 mg group. The permeability index for the control upper limb muscle is shown to the right of each ischemic muscle histogram. Reprinted with permission from Lindsay, TF.* The Albumen Leak after Lower Torso Ischemia Reperfusion. Ann Surg 1992; 216.

than 50% reduction in muscle permeability index measured at 4 hours. In animals treated with sCR1, there was no reduction in the wet to dry ratio of skeletal muscle. No reduction in lung MPO activity was observed at any dose of sCR1. However, at the 6 mg per animal dose, a significant reduction in the lung permeability index was observed. The lytic assays demonstrated that the sCR1 reduced the lytic activity of the classical pathway to near zero for the duration of the experiment. The results of these experiments demonstrate that activation of the complement cascade appears to be an important mechanism of injury resulting in increased muscle permeability to albumin. With regard to the lung, it appears that the change in microvascular permeability in this model requires both the sequestration of neutrophils in the lung via a complement-independent process, while a second complement-dependent step results in increased lung microvascular permeability.

Previous investigators have attempted to reduce injury to the lung following skeletal muscle I/R in sheep and mice.[31] Cobra venom factor was used to deplete complement in sheep and in these animals no significant reduction in lung leukosequestration was noted. Bilateral tourniquet ischemia was created in mice that were genetically deficient in C5 and they had a similar accumulation of neutrophils in their lungs. It was concluded from these studies that pulmonary leukosequestration and edema was not mediated by the complement system but edema was not measured in the C5 deficient mice. Although there are important differences between this study and our own data, it is clear that neutrophils occupy a central role in the pathogenesis of pulmonary dysfunction following remote I/R injury.

SUMMARY

This chapter has reviewed the evidence for skeletal muscle injury by the complement system following I/R. Early studies documented that the washed pellet of homogenized skeletal muscle cells could result in alternate pathway activation. Further studies looking at the anaphylatoxin levels in patients with ischemic limbs demonstrated that there was a significant reduction in the anaphylatoxin levels following amputation. These studies suggested that the complement pathways were activated in patients with ongoing skeletal muscle ischemia. Several studies have pointed to a role for complement in ischemic myocardial injury. Myocardial necrosis was reduced when the soluble complement type 1 (sCR1) receptor was administered.

Recent investigations have begun to shed light on the role of complement in skeletal muscle I/R injury. In a canine model, activation of the alternate pathway was noted during early reperfusion. In addition, the prolonged sequestration of neutrophils following reperfusion of canine skeletal muscle was documented. In animal experiments using sCR1, the permeability index was used as a measure of muscle injury, and the administration of sCR1 resulted in a significant dose dependent reduction in muscle permeability index. However, sCR1 did not effect changes in wet to dry weight. These investigations lend continued support to the hypothesis that complement is a significant mediator of skeletal muscle injury following I/R and further experiments will be required to document a reduction in skeletal muscle necrosis following inhibition of the complement system.

REFERENCES

1. Entman ML, Michael LH, Rossen RD et al. Inflammation in the course of early myocardial ischemia. FASEB J 1991;5:2529-37.
2. Lewis SM, Whatley RE, Cain P et al. Hydrogen peroxide stimulates the synthesis of platelet-activating factor by endothelium and induces endothelial cell-dependent neutrophil adhesion. J Clin Invest 1988; 82:2045-55.
3. Muller-Eberhard HJ. Molecular organization and function of the complement system. Ann Rev Biochem 1988;57:321-47.
4. Goldman G, Welbourn R, Paterson JM et al. Ischemia-induced neutrophil activation and diapedesis is lipoxygenase dependent. Surgery 1990;107:428-33.
5. Von Andrian GH, Hansell P, Chambers JD et al. L-selectin function is required for B_2 integrin mediated neutrophil adhesion at physiologic shear rates in vivo. Am J Physiol 1992;263:H1034-44.
6. Lawrence MB, Springer TA. Leukocytes roll on a selectin at physiologic flow rates: Distinction from and prerequisite for adhesion through integrins. Cell 1991;65:859-73.
7. Dore M, Korthuis RJ, Granger DN et al. P-selectin mediates spontaneous leukocyte rolling in vivo. Blood 1993; 82: 1308-16.
8. Butcher EC. Leukocyte-endothelial cell recognition three (or more) steps to specificity and diversity. Cell 1991; 67:1033-36.
9. Youker K, Smith CW, Anderson D et al. Neutrophil adherence to isolated adult cardiac myocytes. J Clin Invest 1992;89:602-9.
10. Weiss SJ. Tissue destruction by neutrophils. N Engl J Med 1989;320:365-76.
11. Muller-Eberhard HJ. Complement: Chemistry and Pathways. In: Gallin JI, Goldstein IM, Snyderman R,(eds), Inflammation: Basic Principles and Clinical Correlates. New York: Raven Press, Ltd., 1988:21-45.
12. Abbas AK, Lichtman AH, Pober JS. The complement system. In: Cellular and Molecular Immunology, Philadelphia: WB Saunders Co., 1991:259-82.
13. Rossen RD, Michael LH, Kagiyama A et al. Mechanism of complement activation after coronary artery occlusion: Evidence that myocardial ischemia in dogs causes release of constituents of myocardial subcellular origin that complex with human C1q in vivo. Cir Res 1988;62:572-84.
14. Kagiyama A, Savage HE, Michael LH et al. Molecular basis of complement activation in ischemic myocardium: Identification of specific molecules of mitochondrial origin that bind human C1q and fix complement. Cir Res 1989;64:607-15.

15. Crawford MH, Grover FL, Kolb WP et al. Complement and neutrophil activation in the pathogenesis of ischemic myocardial injury. Circulation 1988;78(6):1449-58.
16. Weisman HF, Bartow T, Leppo MK et al. Soluble human complement receptor type 1: In vivo inhibitor of complement suppressing postischemic myocardial inflammation and necrosis. Science 1990;249:146-51.
17. Heideman M, Kaijser B, Gelin LE. Complement activation by homogenized muscle tissue. J Surg Res 1978;25:518-22.
18. Bengtson A, Holmberg P, Heideman M. The ischemic leg as a source of complement activation. Br J Surg 1987;74:697-700.
19. Carden DL, Smith JK, Korthuis RJ. Neutrophil-mediated microvascular dysfunction in postischemic skeletal muscle. Circ Res 1990;66:1436-44.
20. Walden DL, McCutchan HJ, Enquist EG et al. Neutrophils accumulate and contribute to skeletal muscle dysfunction after ischemia-reperfusion. Am J Physiol 1990; 259:H1809-12.
21. Korthuis RJ, Grisham MB, Granger DN. Leukocyte depletion attenuates vascular injury in postischemic skeletal muscle. Am J Physiol 1988;254:H823-27.
22. Yokota J Minei JP, Fantini GA et al. Role of leukocytes in reperfusion injury of skeletal muscle after partial ischemia. Am J Physiol 1989;257:H1068-75.
23. Belkin M, LaMorte ML, Wright JG et al. The role of leukocytes in the pathophysiology of skeletal muscle ischemic injury. J Vasc Surg 1989;10:14-17.
24. Freischlag JA, Hanna D. Neutrophil (PMN) phagocytosis and chemotaxis after 2 hr of ischemia. J Surg Res 1991;50:648-52.
25. Weselcouch EO, Grove RI, Demusz CD et al. Effect of in vivo inhibition of neutrophil adherence on skeletal muscle function during ischemia in ferrets. Am J Physiol 1991;261:H1178-83.
26. Romson JL, Hook BG, Kunkel SL et al. Reduction of the extent of ischemic myocardial injury by neutrophil depletion in the dog. Circulation 1983;67:1016-23.
27. Ma X-L, Lefer DJ, Lefer AM et al. Coronary endothelial and cardiac protective effects of a monoclonal antibody to intercellular adhesion molecule-1 in myocardial ischemia and reperfusion. Circulation 1992; 86: 937-46.
28. Carden DL, Smith JK, Korthuis RJ. Oxidant-mediated CD18-dependent microvascular dysfunction induced by complement-activated granulocytes. Am J Physiol 1991;260:H114-52.
29. Rubin BB, Smith A, Liauw S et al. Complement activation and white cell sequestration in postischemic skeletal muscle. Am J Physiol 1990;259:H525-31.
30. Lindsay TF, Hill J, Ortiz F et al. Blockade of complement activation prevents local and pulmonary albumin leak after lower torso ischemia-reperfusion. Ann Surg 1992; 216(6):677-83.
31. Klausner JM, Paterson IS, Kobzik L et al. Leukotrienes but not complement mediate limb ischemia-induced lung injury. Ann Surg 1989; 209(4):462-470.

CHAPTER 6

Tumor Necrosis Factor and Interleukin-1

Mark Gennaro

Enrico Ascer

Notwithstanding the progress in vascular surgery over the previous 40 years such that surgeons routinely and safely reconstruct diseased vessels and revascularize ischemic tissue in virtually all parts of the body, more recently, interest in the basic pathophysiologic mechanisms causing many complex cellular events has enabled an understanding of the underlying fundamental processes. The intense scrutiny and investigation focused on cellular and subcellular mechanisms has yielded vital information concerning the role of various cell types, molecular and protein messengers, and metabolic pathways in mediating many diverse pathologic processes. One interesting development has been the identification of paracrine protein messengers (cytokines) which appear responsible for intercellular communication and cellular metabolic regulatory functions. The apparent ubiquitous nature of these regulatory proteins and their probable vital role mediating local cellular effects have implicated them as potentially critical elements in the various pathways activated during ischemia-reperfusion. However, despite the wealth of knowledge accumulating regarding the actions of specific cytokines, it must be emphasized that much remains uncertain and further investigation will be required to define their effects in mediating the ischemia-reperfusion syndrome. This chapter will review the current information on tumor necrosis factor alpha (TNFα) and interleukin (IL)-1 regarding their role in ischemia-reperfusion, with particular attention to their effect on endothelial cells and leukocytes.

Ischemia-Reperfusion Injury of Skeletal Muscle, edited by Gary A. Fantini, M.D.;

BASIC INFORMATION

A wide variety of essential cellular activities and functions require communication between cells. One way for this communication to take place is via mediators, such as cytokines and interleukins. These soluble proteins, ranging in molecular weight from 8 to 30,000 daltons, typically are active in minute picomolar-femtomolar concentrations and can exert a wide range of biological effects on different cell types. In addition, different cytokines can act on a particular cell mediating the same effects. The reason for this is that many cytokine receptors are composed of two separate polypeptide moieties: a ligand-binding receptor and a nonbinding signal transducer. The same signal transducer can be activated by different cytokine ligand-binding molecules, explaining how multiple cytokines can cause the same cellular effects. Conversely, as already mentioned, each cytokine can be responsible for a range of biologic actions. For example, IL-1 is an endogenous pyrogen,[1,2] causes the release of corticotropin-releasing factor and adrenocorticotropin in the central nervous system leading to steroid production by the adrenal glands,[3,4] induces prostaglandin PGE_2 and PGI_2 synthesis in both smooth muscle and endothelial cells,[5,6] activates cells of the immune system inducing the production of other cytokines such as TNF and IL-6[7-9] and by inducing the cytotoxicity of monocytes and macrophages against tumor cell lines,[7,10-12] and, in conjunction with IL-3, promotes the growth of primitive stem cells.[13] Table 6.1 is a partial list of the known cytokines along with their cell sources. In a similar fashion, those cytokines which have multiple biologic actions are termed interleukins, while others such as endothelial and fibroblast growth factor have a primary action producing cellular growth. Finally, those cytokines performing primarily antiviral activity are named interferons. Table 6.2 lists some of the functions and activities of selected cytokines. Evident from the above discussion is that most cytokines are produced in response to an external stimulus such as a disease process or after immunization, can elicit measurable tissue responses at very low concentrations, potentiate the release of other cytokines and mediators in complex and often interrelated pathways, contribute to immune responses, inflammation, and endothelial cell activation or are involved in tissue reparative processes.

In addition to the local effects on tissues exerted by cytokines, when produced in excess and circulated systemically, they can significantly alter normal physiologic homeostasis and cause hemodynamic instability, fever, coagulopathy, and organ failure. Though this constellation of findings can be seen in patients with gram negative septicemia, it also has been noted following trauma, myocardial ischemia, and severe pancreatitis.[14-19] In addition, we have previously demonstrated that at least one cytokine, IL-1, is elevated after severe skeletal muscle ischemia-reperfusion, implicating it as a mediator in the reperfusion syndrome which can occur following severe, prolonged ischemia to significant muscle mass.[20] These data and other information regarding the effects of various cytokines on leukocytes and endothelial cells to be discussed below, suggest their importance in producing ischemic tissue injury as well as mediating the systemic effects of the reperfusion syndrome.

CYTOKINES AND LEUKOCYTES

Several studies have provided evidence for the enhancing effects of cytokines on neutrophils. Early work by Klempner et al[21] demonstrated that extremely pure human endogenous pyrogen (IL-1) caused the release of lysozyme and lactoferrin from isolated neutrophils and stimulated oxygen-dependent metabolism in neutrophils.[22] Furthermore, when rabbits were injected with the pyrogen, leukocytes isolated from these animals exhibited increased superoxide anion formation. Bevilacqua and colleagues have reported that IL-1 induces increased neutrophil adherence to cultured human vascular endothelium, via an endothelial cell related mechanism.[23] In addition to activating neutrophils, IL-1 also appears to have chemotactic effects. Purified extracts of IL-1[24-26] and recombinant IL-1 have been noted to stimulate neutrophil chemotaxis.[27,28]

Table 6.1. Partial list of known cytokines

Cytokine	Other Names	Cell Sources
Tumor necrosis factor-α	Cachectin	Blood monocytes Peritoneal macrophages Kupffer's cells Mast cells Endothelial cells Pulmonary macrophages Natural killer cells Glial cells Lymphocytes
Interleukin-1 Interleukin-1α Interleukin-1β	Lymphocyte-activating factor Endogenous pyrogen Osteoclast-activating factor B cell-stimulating factor Catabolin Proteolysis-inducing factor Hemopoietin-1 Leukocyte endogenous mediator	Blood monocytes Peritoneal macrophages Kupffer's cells Endothelial cells Keratinocytes Neutrophils Lymphocytes Epithelial cells Hematopoetic cells
Interleukin-6	Interferon beta2 Hepatocyte-stimulating factor B cell differentiation factor Hybridoma growth factor Hybridoma growth factor B cell-stimulating factor 2	Blood monocytes Endothelial cells Keratinocytes Fibroblassts T cells T cells Macrophages Epithelial cells
Interleukin-2	T cell growth factor	T cells
Interferon gamma	Immune interferon Macrophage-activating factor	Lymphocytes Pulmonary macrophages
Macrophage inflammatory proteins MIP-1 MIP-2	Heparin binding protein	Blood monocytes Peritoneal macrophages
Colony-stimulating factors (CSFs) Interleukin-3	Multipotential CSF Hematopoietic growth factor	T cells Myelomonocytic cells
Granulocyte CSF	Pluripoietin	Macrophages Fibroblasts Granulocytes
Macrophage CSF	CSF-1	Endothelial cells Macrophages Granulocytes
Granulocyte-macrophage CSF	CSF-2	Lymphocytes Endothelial cells Macrophages Fibroblasts Granulocytes

Table 2. Biologic activities of select cytokines

	Tumor Necrosis Factor	Interleukin-1	Interleukin-2	Interleukin-6	Interferon gamma
Immune system Neutrophils	↑Bone marrow release ↑Margination ↑Transendothelial passage ↑Activation	↑Bone marrow release ↑Influx to site of injury ↑Transendothelial passage	Leukocytosis		↑Activation
Monocytes	↑Blood monocyte differentiation ↑Activation ↑Cytotoxicity	↑BM differentiation ↑Activation			↑Cytoxicity ↑Respiratory burst ↑Activation
Lymphocytes	↑Lymphokine production	↑T cell activation ↑Lymphokine production	↑T cell proliferation ↑Cytotoxicity ↑Lymphokine synthesis	↑Differentiation ↑B cell proliferation ↑Cytotoxicity	↑Proliferation ↑B cell antibody production
Eosinophils	Activation of eosinophils		Eosinophilia Organ infiltration		
Cardiovascular system	Hypotenison Shock ↑Vascular leak	?Hypotension	↑Vascular leak ?Hypotension		
Metabolic effects	Anorexia Weight loss Fever	Anorexia Weight loss Fever	?Fever	Fever	
Hepatic	↑Lipogenesis ↑Amino acid uptake ↑Acute-phase protein synthesis ↓Albumin synthesis	↑Acute-phase protein synthesis ↓Albumin synthesis		↑Acute-phase protein synthesis	
Skeletal muscle	↓Resting membrane potential ↑Amino acid loss ↑Myofibrillar protein mRNA ↑Hexose transporters ↓Cellular glycogen ↑Lactate production	↑Nitrogen loss ↓Myofibrillar protein mRNA			
Lipid	Inhibition of lipoprotein lipase ↓Free fatty acid synthesis ↑Lipolysis	Inhibition of lipoprotein lipase			
Effects on wound healing Endothelium	↑Cell surface antigens ↑Procoagulant activity ↑Vascular permeability ↑Vascular proliferation	↑Procoagulant activity Activation of endothelium			
Connective tissue	↑Cartilage resorption ↑Fibroblasdt cytokine production ↑Collagen synthesis ↑Collagenase synthesis	↑Coolagenase activity ↑Proteoglycan degradation ↑Collagen synthesis			
Bone	↑Bone resorption	↑Osteoclast activity			
Fibroblasts				↑Antiviral activity	↑Antiviral activity

Other cytokines exert significant effects on neutrophils as well. Tumor necrosis factor acts on neutrophils to increase the expression of cell surface adherence proteins and C3bi receptors,[29] leading to enhanced neutrophil-endothelial cell interaction. Neutrophil activation with increased production of superoxide radical formation, phagocytosis, and antibody-dependent cytotoxicity have also been attributed to TNF.[30] Interleukin-8, produced by monocytes, fibroblasts, and endothelial cells after induction by TNF and IL-1, similarly activates polymorphonuclear leukocytes and stimulates their production of oxygen-free radicals and release of proteolytic enzymes.[31]

These reports provide indirect support for the role of cytokines in the activation of leukocytes after the perturbation of the cellular environment by ischemia and reperfusion. Many other experimental studies have noted the importance of white blood cells in producing the skeletal muscle injury occurring after ischemia and reperfusion.[32-34] Thus, the hypothesis that cytokines may be responsible for direct effects on ischemic cells, as well as mediate the activation of leukocytes, cellular elements known to participate in tissue injury following ischemia, becomes an attractive one. We have previously demonstrated elevated levels of IL-1 in the venous effluent of skeletal muscle rendered ischemic[20] and it would seem plausible that they mediate activation of these cellular elements. Interestingly, elevation of TNF levels was not found in the canine hind limb model of ischemia and reperfusion.[35] However, this does not discount the role of TNF as a mediator of skeletal muscle ischemia-reperfusion injury, as the multiple interacting metabolic and immunologic pathways governing the regulation of such are undoubtedly complex and will require further investigation to elucidate the exact role each plays. However, because of the similarity of the response noted after the reperfusion of severely ischemic skeletal muscle and that after gram negative sepsis or significant trauma, coupled with the known involvement of cytokines in mediating the biologic effects of these latter two clinical syndromes, the part that cytokines also play in the activation of leukocytes after ischemia-reperfusion is probably integral.

CYTOKINES AND ENDOTHELIAL CELLS

As already noted, tissue ischemia-reperfusion results in a myriad of cellular and subcellular events. One of the important consequences of ischemia is an increase in vascular permeability, clinically manifested as edema. These permeability changes are in turn the result of alterations in the normal integrity of the endothelial cells within the ischemic tissue. The actual mechanism of the permeability changes secondary to endothelial injury have been reported to be caused by oxygen free radicals[36,37] and augmented by activated neutrophils.[38,39] Others have noted the protective effects of the prostacyclin analogue, iloprost, on vascular permeability changes occurring after ischemia-reperfusion and hypothesize that endothelial cell contraction with opening of gap junctions may be a mechanism of increased vascular permeability.[40] Although the exact mechanism or mechanisms of increased vascular endothelial permeability is unclear, it is probable that several pathways contribute to the endothelial dysfunction clinically recognized as edema.

Cytokines have been implicated in vascular permeability changes as well, and may be the mediators of the effects elaborated above. The permeability changes are probably secondary to alteration of the normal morphologic state of the endothelial cell as a result of the action of TNF. Stolpen and coworkers treated human endothelial cells grown to confluence with recombinant TNF and noted elongation and overlap of cells, as well as rearrangement of their actin filaments and loss of fibronectin substratum.[41] Brett and colleagues used cultured bovine aortic endothelial cells treated with recombinant TNF and demonstrated increased permeability to macromolecules and solutes which was reversible.[42] The mechanism appears to result from a change in the disposition of the actin-based cytoskeleton accompanied by cell retraction. These endothelial cell morphologic changes could contribute to the sustained vascular leakage observed after ischemia and reperfusion. However, despite an increase in vascular permeability noted after incubation of cultured bovine aortic endothelial cells with both TNF and IL-1α, Royall et al did not observe an obvious difference in morphology between endothelial cells incubated with either recombinant human TNF or IL-1α and control endothelial monolayers.[43] Although the exact mechanistic pathway causing the permeability changes noted in endothelial cells after incubation with TNF or IL-1 remains in question, these two cytokines directly affect endothelial cells promoting leakage of macromolecules and fluid.

Both TNF and IL-1 can alter the normal hemostatic properties of endothelium causing a procoagulant state. Recombinant TNF incubated with cultured bovine aortic endothelial cells produced a dose dependent rise in procoagulant activity, which was identified as tissue factor by pretreatment with a negating dose of anti-tissue factor IgG.[44,45] Procoagulant activity was noted two hours after TNF treatment with peak activity at ten hours followed by a decrease. This procoagulant effect was the result of de novo protein synthesis as cycloheximide prevented the appearance of the tissue factor. Similarly, IL-1 can increase the expression and synthesis of procoagulant activity both in vivo[46] and in vitro.[47] Local concentrations of these mediators may significantly influence the procoagulant effect since in vitro their effects are dose-dependent and reversible.

Other direct effects of cytokines on endothelial cells are known. Slungaard and colleagues found that human umbilical vein endothelial cells treated with TNF demonstrated increased vulnerability to destruction by the oxidant H_2O_2 and activated eosinophils.[48] Sato and coworkers noted TNF-mediated endothelial cell growth inhibition.[49] Others have reported the stimulatory effect of cytokines on endothelial cells to produce a variety of growth factors and mediators. Zsebo et al[50] noted human umbilical vein endothelial cells incubated with recombinant human IL-1 produced both granulocyte and granulocyte/macrophage colony stimulating factor (G-CSF, G/M-CSF). Sironi and colleagues[51] reported that cultured human umbilical vein endothelial cells stimulated with recombinant IL-1β resulted in markedly increased production of IL-6, a major stimulus of the acute phase response and activator of neutrophils priming them for their respiratory burst and free radical release.[52] Other factors and mediators secreted by cytokine stimulated endothelial cells include IL-1,[53] platelet activating factor (PAF),[54] endothelium-derived relaxing factor,[55] prostacyclin,[56,57] and plasminogen activator inhibitor.[58,59]

As noted above, leukocytes are critical cells producing tissue injury occurring after ischemia and reperfusion. The mechanism(s) leading to their adherence, activation, infiltration, and release of proteolytic enzymes are complex but recent experimental studies have implicated cytokines as fundamental components of this process. Tumor necrosis factor and IL-1 can mediate the adherence of leukocytes to endothelial cells via the induction of expression of endothelial cell surface membrane receptor molecule complexes. At least five distinct human endothelial leukocyte adhesion molecules have been identified (ELAM-1, GMP-140, VCAM-1, ICAM-1, ICAM-2)[60] and the expression of class I major histocompatibility antigens on endothelial cell surfaces has been shown to be stimulated by cytokines as well.[61] For example, endothelial leukocyte adhesion molecule-1 (ELAM-1) is an endothelial surface antigen inducible by TNF and IL-1, initially identified by the binding of mouse monoclonal antibody (mAb) H4/18 to cultured endothelial cells.[62] Basal expression of ELAM-1 by cultured endothelial cells does not occur but rather appears to require the action of either TNF or IL-1. Within 1 hour after treatment with either of these two cytokines, ELAM-1 mRNA appears with cell surface expression following shortly afterwards.[63] After approximately 2 hours, ELAM-1 mRNA levels decline while ELAM-1 cell surface protein expression peaks at 4–6 hours, before diminishing through a process apparently involving endocytosis since ELAM-1 is detectable in permeabilized cells. Interestingly, IL-1 and TNF stimulated neutrophil migration through cultured endothelial cells can involve a transendothelial pathway distinct from the intercellular route typically seen with chemotactic mediators.[64] That this phenomenon is endothelium-dependent is evident by its dependence on de novo protein synthesis and susceptibility to trypsin. Given the importance of endothelial adhesion molecules for infiltration of leukocytes into injured or inflamed tissues and the known role of leukocytes in causing tissue damage in these situations, it is not improbable that cytokines are the fundamental mediators of this response and modulation of their actions will ameliorate any local or systemic deleterious effects.

SUMMARY

The preceding discussion of the role of cytokines, particularly TNF and IL-1, with regard to modulation of ischemia-reperfusion injury in skeletal muscle, has been based largely on both in vivo and in vitro experimental data. And, although much of the investigation illuminating the multiple actions of cytokines has focused on their known association with sepsis and inflammatory processes, their relevance in mediating the effects of skeletal muscle ischemia-reperfusion appears evident, particularly in light of their important actions on leukocytes and endothelial cells. These two cellular elements are integral components within the many interrelated pathways causing tissue injury after ischemia and reperfusion, and the effects of cytokines on both leukocytes and endothelial cells seen in sepsis or inflammatory processes provide compelling indirect evidence that they are prominent mediators of ischemia-reperfusion processes as well. Attempts at modulating the tissue response to ischemia-reperfusion will involve interrupting the actions of cytokines on leukocytes and endothelial cells, perhaps via specific antibodies, thus limiting the subsequent tissue injury and organ damage, and possibly averting cell death.

REFERENCES

1. Dinarello CA, Wolff SM. Molecular basis of fever in humans. Am J Med 1982, 72:799-819.
2. Dinarello CA. Interleukin-1. Rev Infect Dis 1984;6:51-95.
3. Besordorsky H, Del Ray A, Sorkin E et al. Immunoregulatory feedback between interleukin-1 and glucocorticoid hormones. Science 1986;233:652-4.
4. Lee SW, Tsou AP, Chan H et al. Glucocorticoids selectively inhibit the transcription of interleukin-1 beta gene and decrease the stability of interleukin-1 beta mRNA. Proc Natl Acad Sci USA 1988;85:1204-8.
5. Dinarello CA. Interleukin-1. Its multiple biological effects and its association with hemodialysis. Blood Purif 1988;6:164-72.
6. Dejana E, Brenario F, Errol A et al. Modulation of endothelial cell function by different molecular species of interleukin-1. Blood 1987;69:635-99.
7. Philip R, Epstein LB. Tumor necrosis factor as immunomodulator and mediator of monocyte cytotoxicity induced by itself, gamma-interferon, and interleukin-1. Nature 1986;323:86-9.
8. Navarro S, Debili N, Bernaudin JF et al. Regulation of the expression of IL-6 in human monocytes. J Immunol 1989; 142:4339-45.
9. Kishimoto T. The biology of interleukin-6. Blood 1989;74:1-10.
10. Onazaki K, Matsushima K, Kleinerman ES et al. Role of interleukin-1 (IL-1) in promoting human mediated tumor cytotoxicity. J Immunol 1985;135:314-20.
11. Lovett D, Kozan B, Hadam M et al. Macrophage cytotoxicity: interleukin-1 as a mediator of tumor cytostasis. J Immunol 1986;136:340-7.
12. Okubo A, Sone S, Tanaka M et al. Membrane-associated interleukin-1 as a mediator of tumor cell killing of human blood monocytes fixed with paraformaldehyde. Cancer Res 1989;49:265-70.
13. Stanley ER, Bartocci A, Patinkin D et al. Regulation of very primitive multipotent hemopoietic cells by hemopoietin-1. Cell 1986;45:667-74.
14. Neta R, Oppenheim JJ. Why should internists be interested in interleukin-1? Ann Intern Med 1988;109:1-3.
15. Neta R, Sayers TJ, Oppenheim JJ. Relationship of TNF to interleukins. Immunol Ser 1992;56:499-566.
16. Maury CP. Tumor necrosis factor: an overview. Acta Med Scand 1986;220:387-94.

17. Movat HZ. Tumor necrosis factor and interleukin-1: role in acute inflammation and microvascular injury. J Lab Clin Med 1987;110:668-81.
18. Nathan CF. Secretory products of macrophages. J Clin Invest 1987;79:319-26.
19. Old LJ. Tumor necrosis factor (TNF). Science 1985;230:630-2.
20. Ascer E, Mohan C, Gennaro M et al. Interleukin-1 and thromboxane release after skeletal muscle ischemia and reperfusion. Ann Vasc Surg 1992;6:69-73.
21. Klempner MS, Dinarello CA, Gallin JI. Human leukocyte pyrogen induces release of granule contents from human neutrophils. J Clin Invest 1978;61:1330-6.
22. Klempner MS, Dinarello CA, Henderson WR et al. Stimulation of neutrophil oxygen-dependent metabolism by human leukocyte pyrogen. J Clin Invest 1979; 64:996-1002.
23. Bevilacqua MP, Pober JS, Wheeler ME et al. Interleukin 1 acts on cultured human vascular endothelium to increase the adhesion of polymorphonuclear leukocytes, monocytes, and related leukocyte cell lines. J Clin Invest 1985;76:2003-11.
24. Cybulsky MI, Colditz IG, Movat HZ. The role of interleukin 1 in neutrophil-leukocyte emigration induced by endotoxin. Am J Pathol 1986;124:367-72.
25. Sauder DN, Mounessa NL, Katz SI et al. Chemotactic cytokines: the role of leukocyte pyrogen and epidermal cell thymocyte-activating factor in neutrophil chemotaxis. J Immunol 1984;132:828-32.
26. Terui T, Kato T, Suzuki R et al. Effect of interleukin 1 on chemotaxis and chemiluminescence of human neutrophils. Tohuku J Exp Med 1986;149:317-22.
27. Cybulsky MI, Movat HZ, Dinarello CA. Role of interleukin 1 and tumor necrosis factor-alpha in acute inflammation. Ann Inst Pasteur Immunol 1987;138:505-12.
28. Wankowicz Z, Megyeri P, Issekutz AC. Synergy between tumour necrosis factor alpha and interleukin 1 in the induction of polymorphonuclear leukocyte migration during inflammation. J Leukoc Biol 1988;43:349-56.
29. Gamble JR, Harlan JM, Klebanoff SJ et al. Stimulation of the adherence of neutrophils to umbilical vein endothelium by human recombinant tumor necrosis factor. Proc Natl Acad Sci USA 1985;82:8667-71.
30. Shalaby MR, Aggarwal BB, Rinderknecht E et al. Activation of human polymorphonuclear neutrophil functions by interferon-gamma and tumor necrosis factors. J Immunol 1985;135:2069-73.
31. Baggiolini M, Dewald B, Waltz A. Interleukin-8 and related cytokines. In: Gallin JI, Goldstein IM, Synderman R, eds. Inflammation: Basic Principles and Clinical Correlates. New York, Raven Press, 1992:247-263.
32. Smith JK, Grisham MB, Granger DN et al. Free radical defense mechanisms and neutrophil infiltration in postischemic skeletal muscle. Am J Physiol 1989;256:H789-93.
33. Cambria RA, Anderson RJ, Dikdan G et al. Leukocyte activation in ischemia-reperfusion injury of skeletal muscle. J Surg Res 1991;51:13-17.
34. Yokota J, Minei JP, Fantini GA et al. Role of leukocytes in reperfusion injury of skeletal muscle after partial ischemia. Am J Physiol 1989;257:H1068-75.
35. Ascer E, Gennaro M, Cupo S et al. Do cytokines play a role in skeletal muscle ischemia and reperfusion? J Cardiovasc Surg 1992;33:588-92.
36. Parks DA, Granger DN. Ischemia-induced vascular changes: role of xanthine oxidase and hydroxyl radicals. Am J Physiol 1983;245:G285-9.
37. Korthuis RJ, Granger DN, Townsley MI et al. The role of oxygen-derived free radicals in ischemia-induced increases in canine skeletal muscle vascular permeability. Circ Res 1985;57:599-609.
38. Del Maestro RF, Thaw HH, Bjork J et al. Free radicals as mediators of tissue injury. Acta Physiol Scand (Suppl) 1980;492:43-57.
39. Del Maestro RF, Planker M, Arfors KE. Evidence for the participation of superoxide anion radical in altering the adhesive interaction between granulocytes and endothelium, in vivo. Int J Microcirc Clin Exp 1982;1:105-20.

40. Mueller B, Schmidtke M, Witt W. Action of the stable prostacyclin analogue iloprost on microvascular tone and permeability in the hamster cheek pouch. Prostag Leukotr Med 1987;29:187-98.
41. Stolpen AH, Guinan EC, Fiers W et al. Recombinant tumor necrosis factor and immune interferon act singly and in combination to reorganize human vascular endothelial cell monolayers. Am J Pathol 1986;123:16-24.
42. Brett J, Gerlach H, Nawroth P et al. Tumor necrosis factor/cachectin increases permeability of endothelial cell monolayers by a mechanism involving regulatory G proteins. J Exp Med 1989;169:1977-91.
43. Royall JA, Berkow RL, Beckman JS et al. Tumor necrosis factor and interleukin 1 alpha increase vascular endothelial permeability. Am J Physiol 1989;257;L399-410.
44. Nawroth PP, Stern DM. Modulation of endothelial cell hemostatic properties by tumor necrosis factor. J Exp Med 1986; 163:740-5.
45. Bevilaqua MP, Pober JS, Majeau GR et al. Recombinant tumor necrosis factor induces procoagulant activity in cultured human vascular endothelium: characterization and comparison with the actions of interleukin-1. Proc Natl Acad Sci USA 1986;83:4533-7.
46. Nawroth PP, Handley DA, Esmon CT et al. Interleukin-1 induces endothelial cell procoagulant while suppressing cell surface anticoagulant activity. Proc Natl Acad Sci USA 1986;83:3460-4.
47. Bevilacqua MP, Pober JS, Wheeler ME et al. Interleukin-1 activation of vascular endothelium. Am J Pathol 1985;121:394-403.
48. Slungaard A, Vercellotti GM, Walker G et al. Tumor necrosis factor α/cachectin stimulates eosinophil oxidant production and toxicity towards human endothelium. J Exp Med 1990;171:2025-41.
49. Sato N, Goto T, Haranaka K et al. Actions of tumor necrosis factor on cultured vascular endothelial cells: morphologic modulation, growth inhibition, and cytotoxicity. J Natl Cancer Inst 1986;76:1113-8.
50. Zsebo KM, Yuschenkoff VN, Schiffer S et al. Vascular endothelial cells and granulopoiesis: interleukin-1 stimulates release of G-CSF and GM-GSF. Blood 1988; 71: 99-103.
51. Sironi M, Breviario F, Proserpio P et al. IL-1 stimulates IL-6 production in endothelial cells. J Immunol 1989;142:549-53.
52. Mullen PG, Windsor AJC, Fisher AA et al. Interleukin-6 primes neutrophils for sustained respiratory burst in vitro. Br J Surg 1992;79:466-7 (abstract).
53. Lockley RM, Heinzel FP, Shepard HM et al. Tumor necrosis factor α and β differ in their capacities to generate interleukin-1 release from human endothelial cells. J Immunol 1987;139:1891-5.
54. Bussolino F, Breviario F, Tetta C et al. Interleukin-1 stimulates platelet activating factor production in cultured human endothelial cells. J Clin Invest 1986;77:2027-33.
55. Dinarello CA, Wolff SM. The role of interleukin-1 in disease. N Engl J Med 1993;328:106-113.
56. Dejana E, Breviario F, Balconi G et al. Stimulation of prostacyclin synthesis in vascular cells by mononuclear cell products. Blood 1984;64:1280-3.
57. Rossi VF, Breviario F, Ghezzi P et al. Prostacyclin synthesis induced in vascular cells by interleukin-1. Science 1985;229;174-6.
58. Emeis JJ, Kooistra T. Interleukin-1 and lipopolysaccharide induce an inhibitor of tissue-type plasminogen activator in vivo and in cultured endothelial cells. J Exp Med 1986;163:1260-6.
59. Nachman RL, Hajjar KA, Silverstein RL et al. Interleukin-1 induces endothelial cell synthesis of plasminogen activator inhibitor. J Exp Med 1986;163;1595-1600.
60. Pober J, Cotran RS. What can be learned from the expression of endothelial adhesion molecules in tissues? Lab Invest 1991; 64:301-5 (editorial).
61. Collins T, Lapierre LA, Fiers W et al. Recombinant human tumor necrosis factor increases mRNA levels and surface expression of HLA-A,B antigens in vascular endothelial cells and dermal fibroblasts in vitro. Proc Natl Acad Sci USA 1986;83:446-50.
62. Pober JS, Bevilacqua MP, Mendrick DL et al. Two distinct monokines, interleukin-1 and tumor necrosis factor each independently induce biosynthesis and transient expression of the same antigen on the surface of cultured human vascular endothelial cells. J Immunol 1986;136:1680-7.

63. Bevilacqua MP, Stengelin PS, Gimbrone MA Jr et al. Endothelial leukocyte adhesion molecule 1: an inducible receptor for neutrophils related to complement regulatory proteins and lectins. Science 1989; 243: 1160-5.

64. Moser R, Schleiffenbaum B, Groscurth P et al. Interleukin-1 and tumor necrosis factor stimulate human vascular endothelial cells to promote transendothelial neutrophil passage. J Clin Invest 1989;83:444-55.

CHAPTER 7

NITRIC OXIDE (ENDOTHELIUM-DERIVED RELAXING FACTOR)

Edith Tzeng

Timothy R. Billiar

Historically, the field of biomedical research had exhibited little interest in nitric oxide (NO) beyond its participation in the destruction of the ozone layer. Over the last six years, however, the impact of this simple gas on many basic physiologic as well as pathophysiologic processes has been realized. Mitchell et al first suggested that nitrogen oxides could be synthesized by mammalian cells in 1916,[1] but further investigation was not pursued until almost six decades later when the study of carcinogenesis ignited interest in nitrosamine metabolism. An early observation that the human gastrointestinal tract produced nitrites and nitrates was reported by Tannenbaum and co-workers in 1978;[2] however, these nitrogenous wastes were credited to enteral bacterial metabolism for years to come. Later studies documenting nitrite and nitrate production by germ-free rats revived the search for an endogenous route for nitrogen oxide synthesis.[3] More interest developed when mammals were determined to excrete elevated levels of urinary nitrogen oxides after exposure to bacterial endotoxin,[4] implicating a potential role for nitrogen oxides during infectious processes. Confirmation of endogenous nitrogen oxide production by a specific cell type came in 1985 when Stuehr and Marletta[5] measured nitrite and nitrate secretion by bacterial endotoxin-activated mammalian macrophages.

A separate group of investigators focusing on the mechanism of murine macrophage cytotoxicity against certain tumor cell lines also uncovered a role for nitrogen oxides. Hibbs et al[6,7] reported L-arginine conversion to

Ischemia-Reperfusion Injury of Skeletal Muscle, edited by Gary A. Fantini, M.D.;

nitrite and nitrate was necessary and sufficient to promote the cytotoxicity associated with activated macrophages, but nitrites and nitrates themselves could not reproduce this cytotoxic effect. Subsequently, authentic NO was shown to mimmick the pattern of macrophage-mediated cell killing.[8] When NO production by activated macrophages was verified,[8-11] NO was deduced to be the active intermediate in arginine metabolism responsible for cell killing. Analogs of L-arginine with substituted guanidino nitrogens block nitrogen oxide synthesis and macrophages grown in the presence of these analogs lose their tumoricidal actions.

Yet another group of investigators was exploring the mechanisms of agonist-dependent vasodilation. In a landmark paper published in 1980, Furchgott and Zawadzki[12] reported that acetylcholine, a potent vasodilator, required an intact functional endothelial cell layer to promote vascular smooth muscle relaxation. They postulated that endothelial cells, in response to stimulation with certain agonists, secrete a labile factor which then mediates smooth muscle relaxation through cGMP production, thus the name endothelium-derived relaxing factor (EDRF). During attempts to characterize and identify EDRF, parallel experiments with authentic NO revealed that NO could elicit similar vasodilation and possessed many of the same biochemical properties as EDRF,[13-16] and endothelial cells had the capacity to synthesize NO.[17] In 1987, it became widely accepted that NO or some intermediate derivative thereof comprised the vasoactive component of EDRF.[13,14,18,19] As intriguing, NO also produced a marked inhibition of platelet aggregation.[20-22] These new found properties of NO were compelling and established the notion of NO as a fundamental physiologic regulatory molecule.

In turn, the field of neuroscience research also ventured into nitrogen oxides in the late 1970s. In the early 1980s, Deguchi reported cGMP formation in the brain required arginine.[23] In 1988, Garthewaite independently reported the release of a labile mediator with pharmacologic properties strikingly similar to NO from rat cerebellar tissue stimulated with N-methyl-D-aspartate (NMDA).[24] Bredt and Snyder also demonstrated NO-mediated enhancement of cGMP levels in glutamate-stimulated cerebellar tissue.[25] Activation of certain nonadrenergic, noncholinergic peripheral neurons initiated the synthesis and release of a similar factor through a synthetic pathway requiring L-arginine as an essential substrate.[26] The centrally- and peripherally-released factors both exhibited smooth muscle relaxing properties and mediated guanylate cyclase activation. In 1989, the link that cGMP production and L-arginine in the brain revolved around the synthesis of NO from L-arginine was discovered.[27] The regulatory functions of NO in the CNS have been suggested to include pattern formation for long-term memory development, mediation of glutamate-induced excitotoxicity of certain neurons, and cell-cell communication.[28-32] With these and the aforementioned discoveries, the diverse and ubiquitous nature of NO was realized and provided support for the multifunctional properties of NO.

These early studies initiated a whole cascade of related and novel investigations. Interest in NO has skyrocketed in the past five years, capturing an audience of basic scientists and academic physicians as well as the biomedical industry. NO has been demonstrated to mediate varied physiologic responses ranging from hemodynamic regulation and intercellular communication to host defense mechanisms and plays an integral part in immunoregulation. Not limited to host-friendly actions, NO has also been implicated in many deleterious effects, including the destruction of islet cells in diabetes mellitus,[33,34] cartilage damage in arthritis,[35-37] and septic shock,[38-40] to name a few. New aspects of NO biology are still coming to the forefront. Since the early work in the 1980s, a whole gamut of cell types in a wide range of species have been identified to synthesize NO. The list is ever growing and includes macrophages, endothelial cells, hepatocytes, smooth muscle cells, neurons, fibroblasts, islets of Langerhans, chondrocytes, and cardiac myocytes. Current efforts are focused on the biosynthetic and regulatory pathways of NO metabolism as well as the interactions of NO with other intracellular macromolecules. These studies are likely to increase our understanding of how this deceivingly simple gas can elicit such a myriad of effects in a diverse distribution of tissues.

BIOSYNTHESIS AND MECHANISM OF ACTION

NO biosynthesis begins with a nitric oxide synthase (NOS) family of enzymes which catalyzes a 5-electron oxidation of one of the guanidino nitrogens of L-arginine in a reaction with molecular oxygen[41,42] to generate NO and its by-product L-citrulline (Fig. 7.1). NADPH is an essential substrate for this reaction,[43,44] serving as an electron donor. These enzymes also require a host of cofactors including FAD, FMN, heme and tetrahydrobiopterin,[45-48] as well as exhibiting variable dependence on calcium/calmodulin binding.[49-52] This pathway can be inhibited competitively by N^G-substituted L-arginine analogs such as L-nitro^G-monomethyl-arginine (L-NMMA). Once produced, NO is extremely labile with a half-life on the order of a few seconds and is rapidly oxidized to the stable, inactive endproducts of nitrite and nitrate through yet uncertain pathways.[53] Given this decay process, the effects of NO are then limited to and by its immediate environment.

NO elicits its physiologic and pathophysiologic actions through mechanisms which are still under intense investigation and include NOS enzyme activity modulation and protein interactions. One known mechanism, as already eluded to, is activation of soluble guanylate cyclase[53,54] with cGMP in target cells serving as the intracellular signal for NO-mediated effects. This method of signal transduction is indeed how NO mediates vasodilation[55,56] and inhibits platelet adherence and aggregation.[57,58] Guanylate cyclase antagonists can abolish these effects while cGMP agonists or inhibitors of cGMP degradation can augment them. Similarly, neurotransmission by NO is mediated through a cGMP intermediate in the postsynaptic neuron.[25,59] This molecule serves as a physiologic regulator but can alternatively manifest cytostatic and cytotoxic potential. These traits may be a function of local NO concentrations. The larger quantities of NO generated by activated macrophages may overwhelm surrounding cells, allowing more NO to diffuse through the plasma and mitochondrial membranes of

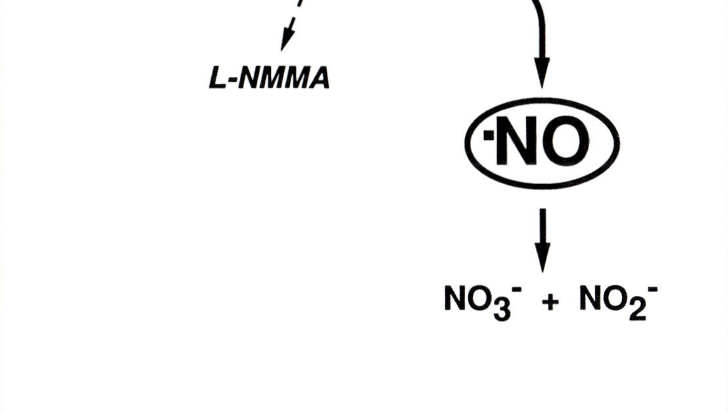

Figure 1: Nitric oxide biosynthetic pathway

Fig. 7.1. Nitric oxide biosynthetic pathway. Nitric oxide synthase (NOS) oxidizes L-arginine with molecular oxygen to nitric oxide (NO) and by-product L-citrulline in a pathway requiring FAD, FMN, heme, and tetrahydrobiopterin (BH_4). Calcium dependence is variable between the isoforms of NOS. L-NMMA can competitively block the formation of NO. NO is further oxidized to end products nitrite (NO_2^-) and nitrate (NO_3^-).

susceptible cells. Intracellularly, NO binds with great affinity to iron-containing enzymes located in the cytosol and the mitochondria, interfering with basic cellular respiration and DNA synthesis[60-64] to arrest cell growth or prompt cell death. Another potential molecular target for NO is superoxide anion. The reaction between NO and superoxide may represent a detoxification pathway for either or both molecules. It may also be the route for the generation of peroxynitrite which can decay to nitrogen dioxide and hydroxyl radical,[65] both potent cytotoxic entities. Finally, the interaction between NO and sulfhydryls has been well-established and has been the basis of food preservation for centuries. Other physiologically relevant molecular interactions with NO are sure to exist but are yet to be characterized. The exact interaction that predominates within specific cell types will determine what effect NO has in each setting, be it advantageous or deleterious, and these interactions can vary moment to moment as influenced by fluctuations in the extracellular/intracellular environment.

A disparity exists between the numerous proposed functions of NO/EDRF and the considerable instability of the molecule. How is it conceivable that this simple gas with a biologic half-life of seconds can mediate the multitude of actions we propose while sizable barriers lie in its path? As already described, NO undergoes rapid oxidation to nitrite and nitrate. It can also be bound and inactivated by hemoglobin and reactive oxygen species. Perhaps there exist extracellular and intracellular carrier molecules which stabilize NO and preserve its biologic activity. Oxides of nitrogen have been shown to react readily with certain cellular proteins, favoring thiol groups on these proteins. The products are S-nitrosothiol adducts which can later serve as NO reservoirs.[66] Keaney et al[67] and Gaston et al[68] have recently demonstrated the presence of S-nitrosothiol species, specifically involving albumin and glutathione. They also showed these S-nitrosothiols possess vasodilator and antiplatelet activities not unlike nitroglycerine and nitroprusside but are active over longer periods of time than authentic NO. These thiols may function as carrier molecules for NO to prolong its bioactivity, help direct site-specific actions at distant targets, and may potentially represent a transcellular shuttling system for NO transport. A role for glyceraldehyde-3-phosphate dehydrogenase in the intracellular shuttling of NO in platelets has been recently suggested.[69]

NITRIC OXIDE SYNTHASES

Currently, three distinct NOS isoforms, as defined by variations in chromosomal localization, subcellular location, amino acid sequence and size, regulation, and function, have been identified. The three isoforms can be segregated into two basic categories; constitutive and inducible. Constitutive NOS (cNOS) can be further subgrouped into the endothelial and the neuronal isoforms while only one inducible NOS has been isolated to date. Chromosomal mapping of all three isoforms in humans has deduced the location of the coding region for endothelial cNOS to chromosome 7,[70] neuronal cNOS to chromosome 12,[70] and inducible NOS (iNOS) to chromosome17,[71] substantiating that each enzyme represents a unique gene product. More tissue-specific NOS isoforms may exist and the search is ongoing. While cNOS expression has been identified in only a few specific tissues thus far, the cellular capacity to express iNOS appears to be more ubiquitous and has been demonstrated in a broader range of tissues (Table 7.1).

Constitutive NOS enzymes are expressed in a continuous manner, as implied by the nomenclature, contributing a low-level basal production of NO. Endothelial cNOS resides primarily on microsomal and plasma membranes[72] while neuronal and inducible NOS are both cytosolic proteins.[44,49] Endothelial cNOS membrane binding is due to an amino terminal myristylation site[73-76] absent from the other isoforms. In addition to requirements for tetrahydrobiopterin, FAD, and FMN, cNOSs demonstrate a calcium and calmodulin dependence. Endothelial cell stimulation by agonists such as acetylcholine or the activation of neuronal NMDA receptors results in increased intracellular calcium which promotes reversible calmodulin binding to NOS;[49,52] this bound calmodulin enhances NOS enzymatic activity, increasing NO production above basal levels. As the calcium flux returns to baseline, the calmodulin moiety dissociates and enzyme activity resumes preexcitation levels. These aspects of cNOS regulation reflect the role of constitutive NO in basic physiologic regulatory functions.

iNOS expression, in contrast, is under complex regulation and is not found in the resting cell. Instead, cells require tissue-specific induction to manifest iNOS expression. The appropriate stimulation includes bacterial invasion, bacterial products, and/or a systemic inflammatory process. Cellular responses to these insults can be mimicked by administering lipopolysaccharide and/or certain cytokines such as TNFα, IFNγ, and IL1.[77-81] Regulation of iNOS expression is linked to the proposed functions for inducible NO. Inducible NO is not believed to be a regulator molecule but is synthesized in response to host perturbations which generate inflammatory or immunologic reactions such as microbial infection and sepsis. Its functions include microbial cytotoxicity as well as host cytoprotection. Uncontrolled activation of the inducible pathway can lead to inadvertent host tissue damage and destruction as presumed to be the underlying etiology for certain autoimmune diseases.

Once iNOS is expressed, NO is generated in a sustained manner for the life of the active enzyme, believed to be on the order of several hours (unpublished data, Billiar et al). It is likely that NO production by iNOS is dependent on substrate and enzyme cofactor availability. Because iNOS is synthesized de novo, maximal NO production is typically delayed for 6–8 hours after induction.[77,79] Tetrahydrobiopterin, an essential cofactor, appears to be rate limiting in NO production by all NOS enzymes.[82-84] The basal level of tetrahydrobiopterin synthesized in a resting cell is adequate to support cNOS activity. However, the induced cell can generate massive amounts of iNOS protein. If the cellular production of tetrahydrobiopterin is not simultaneously increased, iNOS activity will be curtailed. In the master plan of the intricate cellular machinery, this relationship is not ignored and tetrahydrobiopterin biosynthetic enzymes are subject to similar upregulation by endotoxin/cytokine stimulation[85,86] to provide enough cofactor to optimize NO production. Tetrahydrobiopterin may serve as an important structural element, binding dimers of iNOS protein

Table 7.1. Human nitric oxide synthase isoforms

Isoform	Chromosome	Tissue	Proposed Functions
endothelial cNOS	7	endothelial cells	vasomotor regulation platelet homeostasis inhibition of smooth muscle proliferation
neuronal cNOS	12	neuronal tissue, skeletal muscle	cell-cell communication memory formation neurotransmission excitotoxicity
iNOS	17	monocytes /macrophages hepatocytes bone marrow keratinocytes cardiac myocytes pancreatic beta islets	antitumor antimicrobial hepatic cytoprotection immune modulation autoimmune injury free radical scavenger

together to derive the active configuration of the enzyme.[87] In its absence, no active enzyme can form despite abundant iNOS protein. The substrate is derived from either arginine uptake which can be increased coincident with iNOS expression[88] and increased arginine biosynthesis.[89] Unlike cNOS, murine macrophage iNOS co-isolates with an irreversibly bound calmodulin moiety,[90] rendering murine macrophage iNOS entirely calcium/calmodulin insensitive. Historically, inducible NOS was considered unique from cNOS by virtue of absence of calcium dependence. However, isolation and characterization of the human hepatic iNOS[91] has offered contrary data. This iNOS, in distinction from the murine isolate, demonstrates a partial dependence on calcium/calmodulin concentrations. The calmodulin moiety of the human iNOS may bind with a lower affinity than in the murine macrophage iNOS and can be influenced by extremes in calcium fluxes. A level of regulation also exists at induction where dexamethasone can decrease iNOS expression in response to activation with cytokines.[77,92,93] Induction can also be inhibited with IL-4, IL-8, IL-10, TGF-β-1, -2 and -3, and macrophage deactivating factor.[94] A final level of regulation involves posttranscriptional and posttranslational elements which can stabilize/destabilize mRNA or the protein itself, thereby influencing the amount of functional iNOS produced. This is a relatively young field and the complexity of iNOS gene expression regulation will demand considerably more deciphering.

iNOS inducibility and function are not uniform as revealed by tremendous interspecies variability as well as tissue-specific variability. For example, the murine macrophage cell line RAW 264.7 responds vigorously to LPS and IFNγ challenges[95-97] but the human counterpart produces markedly less NO despite prolonged exposure to GM-CSF plus TNF, IFNγ plus TNF, or LPS in culture.[98-100] In contrast, human hepatocytes can be stimulated to exhibit the same high iNOS activity as rat hepatocytes using similar combinations of IL-1, TNF, IFNγ, and LPS.[81] In spite of these differences, comparison of the human hepatic iNOS to the murine macrophage iNOS reveals a shared homology of 80% at both amino acid and DNA sequences while each only sharing 50% homology with any known cNOS sequence.[91] Taking into account species variations, iNOS is well conserved and may represent a single gene product that has evolved tissue specific expression in response to evolutionary pressures. The ease with which hepatocytes can be induced to generate massive quantities of NO may reflect the beneficial role NO plays in combatting malarial and parasitic infections of the liver.[101-103] High levels of NO, on the other hand, have been shown to be toxic to human macrophages which may explain the higher threshold for iNOS expression in this cell type.

NO AS THE ENDOTHELIUM-DERIVED RELAXING FACTOR

Prior to the discovery of EDRF, nitrogen-oxide derived vasodilator drugs including nitroglycerine and nitroprusside have long been utilized clinically to pharmacologically manipulate systemic vascular resistance and blood pressure. Comprehension of the precise mechanisms behind the actions of these drugs lagged many years behind their clinical application. These nitroso compounds were later found to work through guanylate cyclase activation in vascular smooth muscle cells and intracellular cGMP accumulation.[104-106] In addition, these compounds react with thiols such as cysteine and glutathione to generate unstable S-nitrosothiol intermediates which decompose to liberate NO,[107] also a guanylate cyclase activator. Further observations indicated these nitroso compounds elicited antiplatelet effects through cGMP.[108-112] Duplication of the studies with authentic NO produced similar vasorelaxant and platelet antiaggregatory effects through a cGMP intermediate. The effects of these nitroso compounds could be obliterated by hemoglobin or other NO-scavenging proteins as well as inhibitors of guanylate cyclase activity.[108,109] This provided confirmatory evidence that the NO released from nitrogen oxide-derived vasodilators is the bioactive component of these medications.

Characterization of EDRF uncovered uncanny similarities between the chemical properties and actions of EDRF and those of authentic NO.[16,18,19,21] Both compounds generate cGMP through guanylate cyclase activation with subsequent vasodilation and inhibition of platelet adhesion. EDRF had a biologic half-life between 3–5 seconds and was destabilized by superoxide anion, explaining the prolongation of EDRF activity by superoxide dismutase and low oxygen tensions.[112] Hemoglobin and methylene blue diminished or eliminated the bioactivity of EDRF.[16,19,21,113] The advent of electon paramagnetic resonance (EPR) spectroscopy allowed the identification of different paramagnetic molecules and could uniquely recognize EDRF-metalloprotein complexes. Authentic NO similarly exhibited chemical lability, inhibition by hemoglobin and methylene blue, and an EPR spectrum indiscernible from EDRF when complexed to hemoglobin.[112,114] Finally, EDRF and ozone produce a chemiluminescent product with properties equivalent to the product of the reaction between NO and ozone. With these compelling data, Furchgott[14] and Ignarro[19] independently suggested in 1986 that NO or some derivative of NO accounted for the biological activity of EDRF. In 1987, Palmer et al identified and quantitated the NO actually produced and released by vascular endothelial cells. Adequate local NO concentrations were measured to promote the degree of smooth muscle relaxation[17,18] attributed to EDRF, providing proof that EDRF was indeed NO.

Vascularly derived NO appears to be intimately involved in the physiologic maintenance of systemic perfusion. Prior to the discovery of EDRF or NO, the predominant regulatory elements in blood pressure homeostasis were thought to be angiotension and norepinephrine, providing a basal level of vasoconstriction. However, it has since been delineated that the systemic vasculature, ranging from large to microvessels, is maintained in a state of active vasodilation by constitutively released NO.[115,116] Endothelial NO production is influenced by sheer stress and pulsatile flow[117,118] and is also enhanced by certain endothelium-dependent vasodilators. Derangements in NO metabolism and synthesis have been implicated in the pathogenesis of essential hypertension.[116,117] Systemic administration of L-NMMA in a variety of species[115,116,121] including humans[122] increases systemic blood pressure and decreases systemic blood flow by approximately 50%. Administration of L-NMMA also reduced splanchnic, coronary, and cerebral blood flow,[123,124] demonstrating the essential role of NO in regulating vascular tone and maintaining adequate organ perfusion during systemic flow perturbations. Secondly, NO is involved in the homeostatic regulation of platelet aggregation and clot formation. Agents such as collagen, adenosine diphosphate, and thrombin increase platelet intracellular calcium[125] and can instigate platelet aggregation. Increasing intracellular calcium, however, also augments cNOS function to increase NO production which activates guanylate cyclase and cGMP elevation. This culminates in an antiaggregatory phenomenon similar to, but much more potent than that mediated by prostacyclin.[20-22] The antiaggregatory effects of NO are synergistic with those of prostacyclin, providing physiologically relevant antiplatelet actions at subtherapeutic levels of both NO and prostacyclin. In this way, NO provides an important feedback mechanism to downregulate the cascade toward inadvertent thrombus formation and maintains vascular patency.

The pathogenesis of atherosclerotic vascular disease and systemic hypertension is multifaceted and cannot be pinpointed to a single etiologic factor. NO availability has been suggested to be a contributing factor in these processes. An early hypothesis presumed NO exerted cytotoxic effects on the vascular endothelium[126] and promoted endothelial dysfunction with immune complex deposition,[127] increased microvascular permeability, and eventual atherosclerotic plaque formation. More recently, numerous studies have pointed to a protective role for NO in vascular disease. Diminished NO activity results in increased vascular tone and produces a state of hypertension.[119-21] Diabetic patients are of great interest because of their propensity to develop an aggressive form of atherosclerotic degeneration of their vasculature and suffer numerous debilitating cardiac, cerebral, and peripheral vascular consequences. EDRF-mediated vasorelaxation has been found to be deficient in blood vessels isolated from genetic[128,129] and drug-induced[130] diabetic rats. Also, advanced glycosylation products, found in abundance in diabetic patients[131-133] and seen to accumulate during the normal aging process,[134,135] have been implicated in the complications suffered by diabetic and aging patients and have been shown to quench NO.[136,137] Atherosclerotic human coronary arteries[138] and vessels isolated from animals with diet-induced atherosclerosis[139,140] produce decreased amounts of EDRF and enhance responses to pharmacologic vasoconstrictors. The ability of coronary vessels isolated from diseased hearts to respond to acetylcholine challenge is also reduced.[141] Finally, pulmonary artery segments from explanted end-stage lungs have impaired endothelium-dependent vasorelaxation.[142-144] All these observations support the protective role NO is likely to play in maintaining vascular integrity. Its absence would predispose to abnormal vasoconstriction or vasospasm as well as increased platelet deposition and smooth muscle proliferation. Platelet deposition and aggregation would aggravate the process by releasing vasoactive substances to further exacerbate vasospasm and diminish tissue perfusion. Canine coronary artery rings denuded of endothelium responded to platelet degranulation by vasoconstricting while rings with intact endothelial surfaces responded with vasodilation.[145] That replenishing endothelial NO may serve to arrest the atherosclerotic process and its consequences provides some of the driving forces behind the escalating efforts in NO research.

NO AND FREE RADICALS

Another area of NO/EDRF biology prompting great curiosity is ischemia and reperfusion induced tissue injury. A large body of the research in this subject has been performed with myocardial reperfusion models. In 1982, Ku[146] reported that canine coronary artery rings isolated from ischemic hearts subjected to reperfusion responded with diminished vasorelaxation to thrombin stimulus. This observation was verified by reports that ischemia/reperfusion reduced coronary vessel responsiveness to a variety of endothelium-dependent vasodilators including acetylcholine, bradykinin, and leukotriene D4.[147-149] Explanations for this phenomenon initially centered on endothelial dysfunction due to direct cellular injury by invading neutrophils releasing cytotoxic agents in the zone of ischemia.[150,151] Careful studies conducted by Lefer et al,[152,153] however, revealed neutrophil invasion into ischemic tissues lagged measurably behind endothelial dysfunction by 2 to 3 hours. Endothelial dysfunction also preceded any visible ultrastructural damage or morphologic changes in the endothelial cells themselves.[152] In addition, coronary vessels isolated from hearts subjected to up to 90 minutes of ischemia retained endothelium-dependent vasodilation to acetylcholine stimulus, while vessels from reperfused postischemic hearts manifested impaired vasorelaxation to the same stimulus as early as 2.5 minutes into reperfusion, but retained the ability to respond to authentic NO.[152,153] These observations support the role of NO/EDRF as a cytoprotective factor in ischemia/reperfusion injury, and lack of NO/EDRF activity may promote a more extensive and irreversible injury.

EDRF stability was noted to be adversely influenced by high oxygen tensions but is protected by superoxide dismutase (SOD).[112,155,156] Reactive oxygen species, namely superoxide anion, are generated by activated neutrophils, monocytes, and macrophages as well as other cell types and can be augmented in the setting of excess oxygen. These free radicals can react with and inactivate NO, also a free radical. SOD reportedly enhances and prolongs NO activity by scavenging superoxide through its conversion to hydrogen peroxide which can then be eliminated by catalase. A direct interaction between NO and Cu,Zn-SOD[157] has not been observed. Superoxide anion and its degradation products are cytotoxic free radicals which are released in large quantities by activated PMNs during states of inflammation or infection as well as following ischemia/reperfusion,[158-160] and are thought to be the etiology of the resulting tissue injury. Superoxide anion can cause lipid peroxidation, which disrupts cell and mitochondrial membranes as well as impairing calcium transport.[163] Because of the affinity between NO and superoxide, a hypothesis was formulated that NO may behave as a natural extracellular scavenger for reactive oxygen species, providing a biologic barrier to the toxicity of these agents. In the myocardial ischemia/reperfusion model, at least two mechanisms may be postulated to account for the observed decrease in NO activity: (1) decreased production of NO due to endothelial cell dysfunction, or (2) scavenging or deactivation of NO by superoxide anion or other free radical species. As previously noted, decreased NO activity was measured prior to neutrophil invasion into the myocardium.[153] Also, superoxide anions could be detected in significant amounts in ischemic myocardial tissue prior to neutrophil invasion.[153] These free radicals appear to arise from the vascular endothelial cells themselves and may be responsible for endothelial injury.[152,161,162] Eventual neutrophil adherence can aggravate the situation with further release of cytotoxic agents, augmenting injury to the ischemic and reperfused region.

The significance of reperfusion to the development of tissue injury following an ischemic insult has been studied extensively in various experimental models. Clinical correlates include patients undergoing open heart surgery necessitating extracorporeal circulation, thrombolytic therapy or angioplasty to reestablish myocardial perfusion after acute infarction, organ transplantation, traumatic limb amputation and replantation, and relief of acute vascular occlusion in the extremities. Logically, prolonged ischemia itself will ultimately result in cell death. More clinically relevant, however, is that a substantial portion of the tissue necrosis that does occur is the direct consequence of reoxygenation of previously ischemic tissues[152,153,163,164] and not of the ischemia alone. During periods of ischemia, molecular oxygen is not a readily available substrate intracellularly, limiting any pathway requiring oxygen such as superoxide generation. Similarly, molecular oxygen is a necessary substrate for NO production which is also curtailed during hypoxia or in tissues where oxygen tensions are diminished such as the venous circulation.[165,166] Xanthine oxidase as a biologic source for superoxide radical generation was identified in 1968,[167] and this enzyme appears to be a physiologically important source in a variety of postischemic settings,[168] as it is ubiquitously distributed throughout the body and across species. Xanthine dehydrogenase is the predominant form of the enzyme, represenating 90% of the total activity in healthy tissues.[169] However, the dehydrogenase form of the enzyme is inactive and cannot catalyze the formation of superoxide anion from molecular oxygen (Fig. 7.2). Xanthine dehydrogenase can be transformed to the active form of xanthine oxidase and the conversion appears to be pronounced and accelerated during ischemia. In rat ileum, near complete conversion to the oxidase form occurs within 10 seconds of nonperfusion.[168] In contrast, there is no appreciable conversion of xanthine dehydrogenase to xanthine oxidase (D to O) for up to 1 hour of ischemia in skeletal muscle,[169] which may contribute to the ability of skeletal muscle to resist ischemia/reperfusion injury relative to other organ systems. Type D to O conversion has, however, been documented in skeletal muscle subjected to 2 hours of ischemia.[170] Simultaneous buildup of hypoxanthine, a suitable oxidizable substrate for xanthine oxidase, occurs through the breakdown of AMP which cannot be recycled to ATP in the absence of oxygen. With reperfusion, oxygen availability now provides adequate substrate for xanthine oxidase to produce superoxide anion. Low-level reactive oxygen free radicals are generated in hearts during ischemia but a burst of oxygen radical formation occurs within moments of reperfusion.[171] Similarly, the substrate for NOS is also available during reperfusion and NO production resumes. The importance of molecular oxygen to the development of cellular injury during reperfusion was confirmed by Lefer et al who reported reperfusion in the setting of 95% N_2 and 5% CO_2 failed to influence endothelium-dependent vasorelaxation of isolated coronary artery segments.[153]

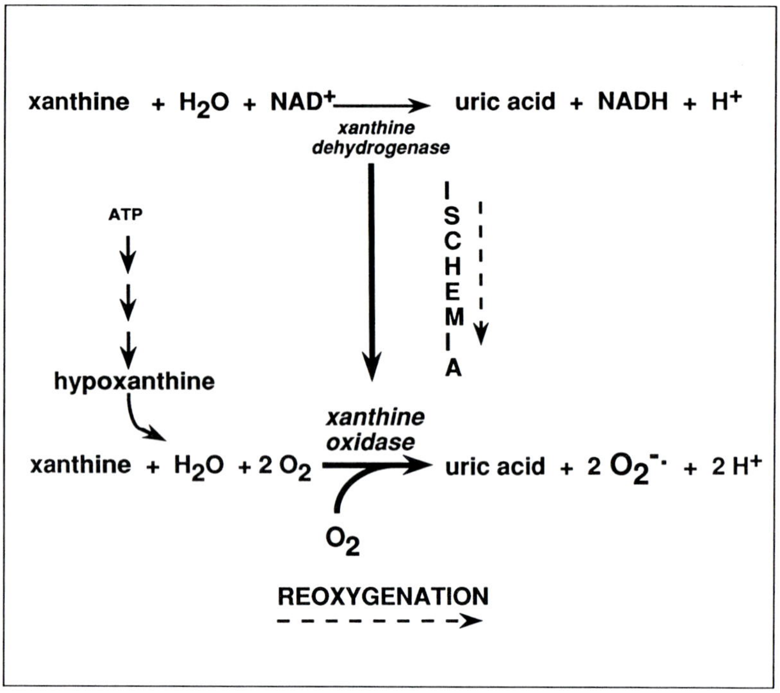

Fig. 7.2. Proposed mechanism of superoxide production in ischemia/reperfusion injury. Xanthine oxidase production from xanthine dehydrogenase is dominant during ischemia. Also produced during ischemia is hypoxanthine, a suitable substrate for xanthine oxidase, from the breakdown of ATP in the absence of oxygen. Upon reoxygenation, substrates for xanthine oxidase are abundant and favors the generation of superoxide anion.

Ischemia promotes cellular changes stemming from inadequate energy production to maintain cellular ion gradients, precipitating a redistribution of calcium ions into the cell. This increased calcium flux can augment endothelial cNOS activity and increase NO production. The increase in NOS activity which accompanies ischemia/reperfusion[166] may again reflect the potential role of NO as a free radical scavanger.[172,173] Further support can be derived from reports where perfusion with calcium free media results in enhanced tissue damage.[174] In addition, there is some evidence to suggest that NO may downregulate xanthine oxidase function[175] which further corroborates the suggested cytoprotective properties of NO. This hypothesis has been evaluated in ischemia/reperfusion models using feline hearts.[164,176] The administration of NO donors reduced the size of infarcted myocardium significantly. The protection afforded by NO was further augmented by SOD as well as other free radical scavangers. Coronary vessel rings from hearts treated with NO donors also retained endothelium-dependent vasorelaxant responses to agonists. Similar cytoprotective effects of authentic NO have been demonstrated in the inactivation of superoxide radicals released by human leukocytes.[172]

The inducible NOS system appears to have a cytoprotective role as well. A well-studied system involves hepatic NO production in response to endotoxin challenge.[77,81,177,178] In the in vivo murine model of liver injury as elicited by *Corynebacterium parvum* infection[177] followed by LPS[179] administration, elevated circulating levels of nitrites and nitrates were generated and severe hepatic damage ensued. Blocking NO production with L-NMMA resulted in a dose-dependent reduction of plasma nitrites and nitrates but an exacerbation of hepatocellular damage as measured by liver transferase levels. The etiology of the hepatocellular damage appears to be the activated hepatic macrophages.[180] These macrophages may induce injury through the release of oxygen free radicals locally which can be scavenged by hepatocyte derived NO. Superoxide dismutase administration abrogates the increased injury induced by NO inhibition,[176] further supporting the role for NO in neutralization of oxygen radicals. The presence of platelet thrombi and microinfarction of the livers of L-NMMA-treated animals has pointed to the importance of the antiplatelet actions of NO in the microcirculation in endotoxemia.[181] The enhancement of liver damage with NOS inhibitors supports this cytoprotective role.

The reaction between NO and superoxide anion produces peroxynitrite ($ONOO^-$) at a rate constant of $3.4 \times 10^7\ M^{-1}\ sec^{-1}$ in aqueous solution.[182] Endothelial cells, neutrophils and macrophages can simultaneously produce NO and superoxide,[173,184] setting the proper microenvironment for the formation of peroxynitrite. Peroxynitrite is a much more potent oxidizing agent than NO[185] or superoxide and has a half-life of approximately one second at physiologic pH,[65] allowing it to diffuse several cell lengths before breakdown occurs. At its target sites, peroxynitrite can react with cell membranes resulting in lipid peroxidation.[186] It can oxidize sulfhydryl moieties 1000-fold faster than hydrogen peroxide.[185] In addition, it can also react with metal ions and with superoxide dismutase to form a potent nitrating agent with the reactivity of nitronium ion.[187,188] Peroxynitrite has been demonstrated to possess bactericidal function with a LD_{50} of 250μ M at 37°C at physiologic pH,[189] suggesting peroxynitrite may be the true cytotoxic mediator or a major contributor in macrophage-mediated cell killing.

In alkaline conditions, peroxynitrite is a stable compound. However, once protonated to peroxynitrous acid, it decays rapidly. At physiologic pH, approximately 20% of peroxynitrite will be protonated to peroxynitrous acid.[185] This compound can decompose in two pathways as dictated by the configuration of the peroxynitrous acid molecule.[65,187] In the cis-configuration, a lower energy and more stable intermediate which predominates in alkaline conditions, peroxynitrous acid decomposes to generate nitrate and a proton, both relatively benign species (Fig. 7.3). The trans isomer is a higher energy configuration which can also decay to a proton and nitrate or can alternatively undergo homolytic decomposition to hydroxyl radical and nitrogen dioxide, both

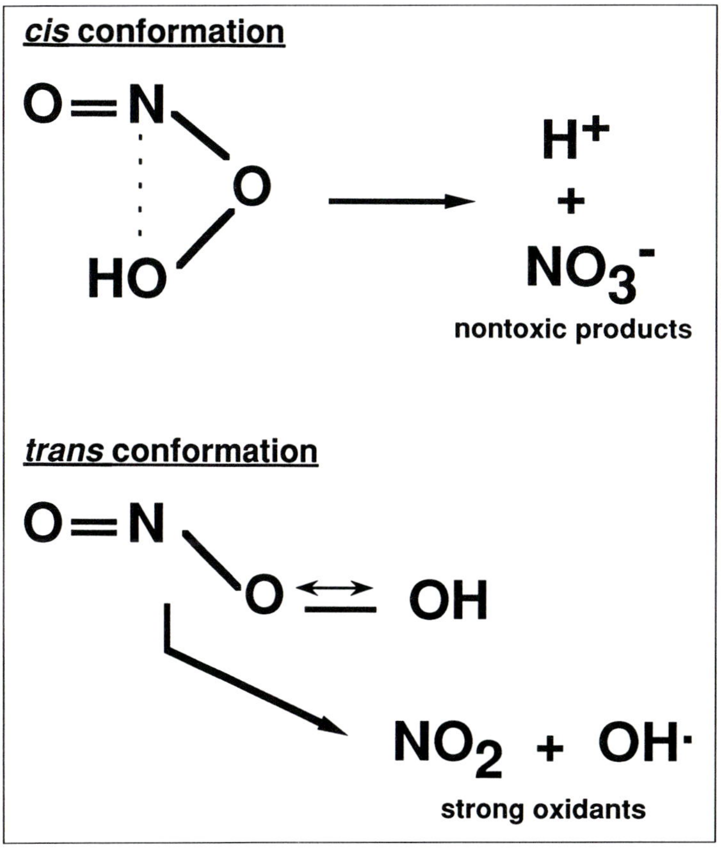

Fig. 7.3. Decomposition of peroxynitrite. When protonated to peroxynitrous acid, the lower energy cis-configuration is favored and nontoxic breakdown products are formed. The higher energy trans-configuration favors homolytic decomposition to the potent oxidants hydroxyl radical ($OH^{\cdot}$) and nitrogen dioxide (NO_2).

extremely potent oxidants. This reaction does not require transition metal catalysis. Free hydroxyl radical and nitrogen dioxide can induce membrane lipid peroxidation with rate constants of 10^9 M^{-1} sec^{-1} for hydroxyl radical[186] and 10^5 M^{-1} s^{-1} for nitrogen dioxide.[190] Clearly, these products of peroxynitrite decomposition have great cytotoxic potential themselves and may also contribute to the tissue injury previously associated with NO and superoxide anion. Peroxynitrite, hydroxyl radical, and nitrogen dioxide may all contribute to the beneficial bactericidal effects of macrophages as well as the detrimental enhanced cellular injury seen with ischemia/reperfusion or autoimuune diseases.

The diverse physiologic and pathophysiologic roles of NO are well reported and encompass both cytoprotective and cytotoxic mechanisms. The question presently being addressed is how this simple molecule can mediate such extremes in biologic function. As already documented, NO participates in the regulation of systemic perfusion, platelet homeostasis, and neurotransmission. On the opposite end of the spectrum, NO can also elicit the toxic effects responsible for its antimicrobial/antitumor actions, tissue injury in ischemia/reperfusion, and the destruction of vital cells such as chondrocytes and pancreatic islets in autoimmune processes. An explanation for these opposing effects may lie in two divergent metabolic pathways (Fig. 7.4). Constitutively produced NO functions predominantly as a physiologic modulator and its effects are primarily tissue conserving. The cytoprotective mechanism involves scavenging free oxygen radicals and generating the less harmful end products of hydrogen ion and nitrate. Another mechanism may focus on hydrogen peroxide. Xanthine oxidase is capable of generating large quantities of superoxide anion, but in the presence of superoxide dismutase, superoxide may be preferentially converted to hydrogen peroxide prior to associating with NO. Hydrogen peroxide is also toxic to cells through metal-mediated reduction to form hydroxyl radical equivalents.[191] There is evidence to suggest that NO can inactivate hydrogen peroxide in a yet to be identified reaction.[173] Finally, xanthine oxidase may be downregulated in the presence of NO,[175] another potential mechanism of NO protection against free oxygen radical-induced cell damage.

The inducible pathway generates massive quantities of NO in response to endotoxin or cytokine stimulation. This NO also serves in a host-protective mechanism by combatting microbial invasion and neoplastic degeneration. The mechanism of this cytotoxicity appears to be through the interaction of NO with free oxygen radicals, namely superoxide anion (Fig. 7.4). While superoxide anion has definite cytotoxic potential, its limited reactivity with biological molecules in vivo raises questions of its true toxicity. The true toxic species may be the products of superoxide metabolism and not superoxide itself. Macrophages, monocytes, and activated neutrophils release NO and superoxide radical simultaneously, providing an environment favoring the production of peroxynitrite. This peroxynitrite and its decomposition products, hydroxyl radical and nitrogen dioxide, then promote the directed cell killing of bacteria, parasites, and tumor cells in the near vicinity. In states of active inflammation and infection, macrophages and neutrophils can release their toxic products in areas of vital host tissue such as the liver during active hepatitis. The hepatocyte is armed with an inducible NOS which can manufacture more NO to scavenge the free radicals. The liver is also an abundant reservoir of SOD which can convert superoxide into hydrogen peroxide to be inactivated by catalase and NO. Inadvertent activation of macrophages and neutrophils through autoimmune processes can result in the release of NO and free oxygen radicals in tissues not equipped for the detoxification of these products, resulting in irreversible host tissue damage. This may be the etiologic modality in the genesis of juvenile diabetes mellitus and rheumatoid arthritis.

Much more information is required to decipher the numerous biochemical reactions in which NO can participate to determine which are biologically relevant. The significance of peroxynitrite formation is still under investigation. The generation of this compound from NO and superoxide anion has been demonstrated quite well in vitro. However, the relevance of peroxynitrite, hydroxyl radical, and nitrogen dioxide in vivo still needs to be established as well as understanding the regulatory role NO may play in xanthine oxidase activity and hydrogen peroxide deactivation. The processes by which NO can mediate both cytoprotective and cytodestructive events are complex and the explanation provided is clearly an oversimplification of what is actually occurring. There are innumerable interrelated reactions that may be significant and may include biochemical pathways yet undiscovered. The road to comprehending the functions of NO is a long and tortuous one with many misleading avenues and the journey has merely begun.

References

1. Mitchell HH, Shonle HA, Grindley HS. The origin of nitrate in the urine. J Biol Chem 1916; 24:461.
2. Tannenbaum SR, Fett D, Young VR et al. Nitrite and nitrate are formed by endogenous synthesis in the human intestine. Science 1978; 200:1487-1489.
3. Green LC, Tannenbaum SR, Goldman P. Nitrate synthesis in the germfree and conventional rat. Science 1981; 212:56-8.

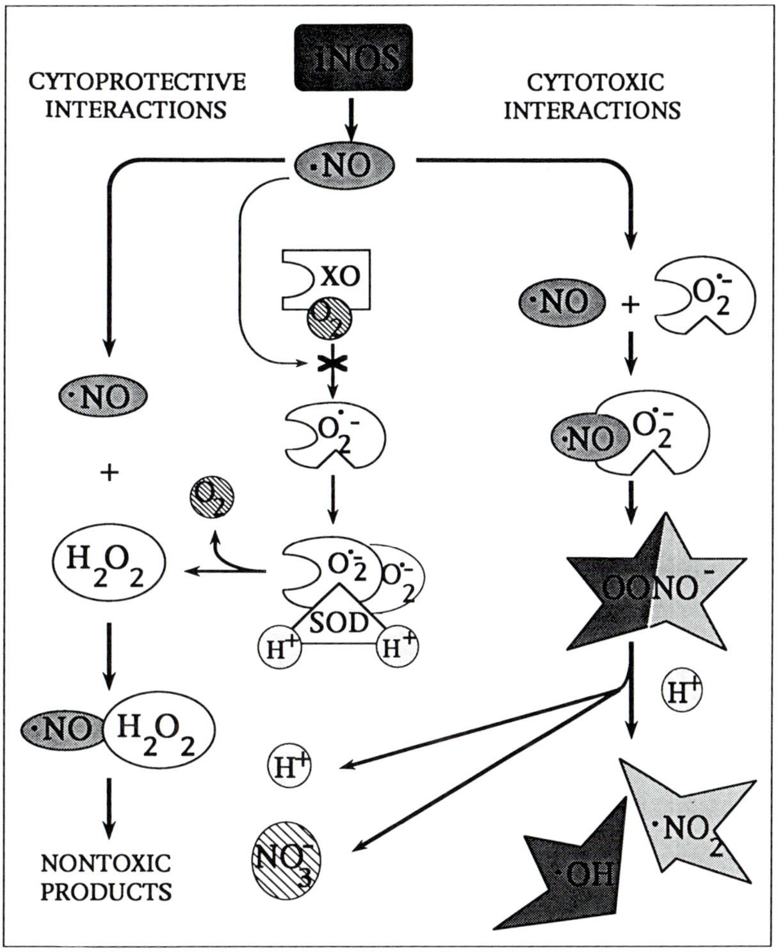

Fig. 7.4. NO interactions with cytotoxic radicals. Possible pathways through which NO mediates its dual functions of cytoprotection and cytotoxicity. NO can react with superoxide anion to produce peroxynitrite which can decompose to more potent cytotoxic agents, nitrogen dioxide and hydroxyl radical. These endproducts may mediate the cytotoxic effects associated with NO and superoxide anion. The decomposition of peroxynitrite can, alternatively, generate benign products, hydrogen ion and nitrate. Superoxide radical can also be scavenged by superoxide dismutase and converted to hydrogen peroxide which may then be deactivated by NO. NO may also downregulate xanthine oxidase production of superoxide anion. The cytoprotective effects of NO may be a combination of these reactions. Abbreviations: inducible nitric oxide synthase = iNOS, nitric oxide = •NO, superoxide anion = $O_2^{-\cdot}$, peroxynitrite = $OONO^-$, hydroxyl radical = •OH, nitrogen dioxide = •NO_2, hydrogen peroxide = H_2O_2, superoxide dismutase = SOD, xanthine oxidase = XO.

4. Wagner DA, Young VR, Tannenbaum SR. Mammalian nitrate biosynthesis: Incorporation of $^{15}NH_3$ into nitrate is enhanced by endotoxin treatment. Proc Natl Acad Sci USA 1983; 80:4518-21.
5. Stuehr DJ, Marletta MA. Mammalian nitrate biosynthesis: mouse macrophages produce nitrite and nitrate in response to *Escherichia coli* lipopolysaccharide. Proc Natl Acad Sci USA 1985; 82:7738-42.
6. Hibbs Jr JB, Taintor RR, Vavrin Z. Macrophage cytotoxicity: role for L-arginine deaminase and imino nitrogen oxidation to nitrate. Science 1987; 235:473-6.
7. Hibbs JB Jr, Vavrin Z, Taintor RR. L-arginine is required for expression of the activated macrophage effector mechanism causing selective metabolic inhibition in target cells. J Immunol 1987; 138:550-65.
8. Hibbs JB Jr, Taintor RR, Vavrin Z et al. Nitric oxide: a cytotoxic activated macrophage effector molecule. Biochem Biophys Res Commun 1988; 157:87-94.
9. Miwa M, Stuehr DJ, Marletta MA et al. Nitrosation of amines by stimulated macrophages. Carcinogenesis 1987; 8:955-8.
10. Marletta MA, Yoon PS, Kyengar R et al. Macrophage oxidation of L-arginine to nitrite and nitrate: nitric oxide is an intermediate. Biochemistry 1988; 27:8706-11.
11. Stuehr DJ, Nathan CF. Nitric oxide: a macrophage produce responsible for cytostasis and respiratory inhibition in tumor target cells. J Exp Med 1989; 169: 1543-55.
12. Furchgott RF, Zawadzki JV. The obligatory role of endothelial cells in the relaxation of arterial smooth muscle by acetylcholine. Nature 1980; 288:373-6.
13. Furchgott RF. Studies on relaxation of rabbit aorta by sodium nitrate: basis for the proposal that the acid-activatable component of the inhibitory factor from retractor penis is inorganic nitrate and the endothelium-derived relaxing factor is nitric oxide. In Vanhoutte PM, ed. Mechanisms of Vasodilatation. New York: Raven, 1988: 401-14.
14. Khan MT, Furchgott RF. Additional evidence that endothelium-derived relaxing factor is nitric oxide. In Rand MJ and Roper C, eds. Pharmacology. Amsterdam: Elsevier, 1987:341-4.
15. Ignarro LJ, Byrns RE, Wood KS. Biochemical and pharmacological properties of endothelium-derived relaxing factor and its similarity to nitric oxide radical. In Vanhoutte PM, eds. Mechanisms of vasodilatation. New York: Raven, 1988:427-35.
16. Ignarro LJ, Byrns RE, Buga GM et al. Endothelium-derived relaxing factor from pulmonary artery and vein possesses pharmacological and chemical properties that are identical to those for nitric oxide radical. Circ Res 61; 1987:866-79.
17. Palmer RMJ, Ashton DS, Moncada S. Vascular endothelial cells synthesize nitric oxide form L-arginine. Nature 1988; 333: 664-6.
18. Palmer RMJ, Ferrige AG, Moncada S. Nitric oxide release accounts form the biological activity of endothelium-derived relaxing factor. Nature 1987; 327:524-5.
19. Ignarro LJ, Buga GM, Wood KS et al. Endothelium-derived relaxing factor (EDRF) produced and released from artery and vein is nitric oxide. Proc Natl Acad Sci USA 1987; 84:9265-9.
20. Radomski MW, Palmer RMJ, Moncada S. The role of nitric oxide and cGMP in platelet adhesion to vascular endothelium. Biochem Biophys Res Commun 1987; 148: 1482-9.
21. Radomski MW, Palmer RMJ, Moncada S. Comparative pharmacology of endothelium-derived relaxing factor, nitric oxide and prostacyclin in platelets. Br J Pharmacol 1987; 92:181-7.
22. Moncada S, Radomski MW, Palmer RMJ. Endothelium-derived relaxing factor: identification as nitric oxide and role in the control of vascular tone and platelet function. Biochem Pharmacol 1988; 37: 2495-501.
23. Deguchi T, Yoshioka M. L-arginine identified as an endogenous activator for soluble guanylate cyclase from neuroblastoma cells. J Biol Chem 1982; 257:10147-52.
24. Garthewaite J, Charles SL, Chess-Williams R. Endothelium-derived relaxing factor release on activation of NMDA receptors suggests role as intercellular messenger in the brain. Nature 1988; 336:385-8.
25. Bredt DS, Snyder SH. Nitric oxide mediates glutamate-linked enhancement of

cGMP levels in the cerebellum. Proc Natl Acad Sci USA 1989; 86:9030-3.
26. Gillespie JS, Liu X, Martin W. The neurotransmitter of the nonadrenergic noncholinergic inhibitory nerves to smooth muscle of the genital system. In Moncada S, Higgs EA, eds. Nitric Oxide From L-araginine: A Bioregulatory System. Amsterdam: Elsevier, 1990:147-64.
27. Knowles RG, Palacios M, Palmer RMJ et al. Formation of nitric oxide from L-arginine in the central nervous system. A transduction mechanism for stimulation of the soluble guanylate cyclase. Proc Natl Acad Sci USA 1989; 86:5159-62.
28. Shibuki K, Okada D. Endogenous nitric oxide release required for long-term synaptic depression in the cerebellum. Nature 1991; 349:326-8.
29. Rothman SM, Olney JW. Excitotoxicity and the NMDA receptor. Trends Neurosci 1987; 10:299-302.
30. Choi DW. Glutamate neurotoxicity and diseases of the nervous system. Neuron 1988; 1:623-34.
31. Garthewaite J. NMDA receptors, neuronal development and neuronal degeneration. In Watkins JC, Collingridge GL, eds. The NMDA Receptor. Oxford: Oxford University Press, 1989:187-205.
32. Bredt DS, Snyder SH. Nitric oxide, a novel neuronal messenger. Neuron 1992; 8:3-11.
33. Corbett JA, Lancaster JR Jr, Sweetland MA et al. Interleukin-1β-induced formation of EPR-detectable iron-nitrosyl complexes in islets of Langerhans: role of nitric oxide in interleukin-1β-induced inhibition of insulin secretion. J Biol Chem 1991; 266: 21351-4.
34. Corbett JA, Wang JL, Hughes JH et al. Nitric oxide and cGMP formation induced by interleukin-1β in islets of Langerhans: evidence for an effector role of nitric oxide in islet dysfunction. Biochem J 1992; 287:229-35.
35. Stefanovic-Racic M, Stadler J, Evans CH. Nitric oxide and arthritis. Arthritis Rheum 1993; 36:1036-44.
36. Stadler J, Stefanovic-Racic M, Billiar TR et al. Articular chondrocytes synthesize nitric oxide in response to cytokines and lipopolysaccharide. J Immunol 1991; 147:3915-20.
37. McCartney-Francis N, Allen JB, Mizel DE et al. Suppression of arthritis by an inhibitor of nitric oxide synthase. J Exp Med 1993; 178:749-54.
38. Kilbourn RG, Jubran A, Gross SS et al. Reversal of endotoxin-mediated shock by N^G-methyl-L-arginine, an inhibitor of nitric oxide synthesis. Biochem Biophys Res Com 1991; 172:1132-8.
39. Kilbourn RG, Gross SS, Jubran A et al. N^G-methyl-L-arginine inhibits tumor necrosis factor-induced hypotension: implications for the involvement of nitric oxide. Proc Natl Acad Sci USA 1990; 84:3629-32.
40. Petros A, Bennett D, Vallance P. Effect of nitric oxide synthase inhibitor on hypotension in patients with septic shock. Lancet 1991; 338:1557-8.
41. Kwon NS, Nathan CF, Gilker C et al. L-citrulline production from L-arginine by macrophage nitric oxide synthase: the ureido oxygen derives from dioxygen. J Biol Chem 1990; 265:13,442-5.
42. Leone AM, Palmer RMJ, Knowles RG et al. Constitutive and inducible nitric oxide synthases incorporate molecular oxygen into both nitric oxide and citrulline. J Biol Chem 1991; 266:23,790-5.
43. Marletta MA, Poksyn SY, Iyengar R et al. Macrophage oxidation of L-arginine to nitrite and nitrate: nitric oxide is an intermediate. Biochemistry 1988; 27:8706-11.
44. Stuehr DJ, Cho HJ, Kwon NS et al. Purification and characterization of the cytokine-induced macrophage nitric oxide synthase: an FAD- and FMN-containing flavoprotein. Proc Natl Acad Sci USA 1991; 88:7773-7.
45. Tayeh MA, Marletta MA. Macrophage oxidation of L-arginine to nitric oxide, nitrite, and nitrate: tetrahydrobiopterin is required as a cofactor. J Biol Chem 1989; 264: 19654-8.
46. Kwon NS, Nathan CF, Stuehr DJ. Reduced biopterin is a cofactor in the generation of nitrogen oxides by murine macrophages. J Biol Chem 1989; 264:20496-501.
47. Stuehr DS, Kwon NS, Nathan CF. FAD and GSH participate in macrophage synthesis of nitric oxide. Biochem Biophys Res Commun 1990; 168:558-65.

48. Mayer B, John M, Heinzel B. Brain nitric oxide synthase is a biopterin- and flavin-containing multi-functional oxido-reductase. FEBS Lett 1991; 288:187-91.
49. Bredt DS, Snyder SH. Isolation of nitric oxide synthase, a calmodulin-requiring enzyme. Proc Natl Acad Sci USA 1990; 87:682-5.
50. Mayer B, Schmidt K, Humbert R et al. Biosynthesis of endothelium-derived relaxing factor: a cytosolic enzyme in porcine aortic endothelial cells Ca^{2+}-dependently converts L-arginine into an activator of soluble guanylyl cyclase. Biochem Biophys Res Commun 1989; 164:678-85.
51. Mulsch A, Bassenge E, Busse R. Nitric oxide synthesis in endothelial cytosol: evidence for a calcium-dependent and a calcium-independent mechanism. Naunyn Schniedebergs Arch Pharmacol 1989; 340:767-70.
52. Busse R, Mulsch A. Calcium-dependent nitric oxide synthesis in endothelial cytosol is mediated by calmodulin. FEBS Lett 1990; 265:133-6.
53. Moncada S, Palmer RMJ, Higgs EA. Nitric oxide: physiology, pathophysiology and pharmacology. Pharmacol Rev 1991; 43:109-42.
54. Ignarro LJ. Signal transduction mechanisms involving nitric oxide. Biochem Pharmacol 1991; 41:485-490.
55. Ignarro LJ, Harbison RG, Wood KS et al. Activation of purified soluble guanylate cyclase by endothelium-derived relaxing factor from intrapulmonary artery and vein: stimulation by acetylcholine, bradykinin and arachidonic acid. J Pharmacol Exp Ther 1986; 237:893-900.
56. Forstermann U, Mulsch A, Bohme E et al. Stimulation of soluble guanylate cyclase by an acetylcholine-induced endothelium-derived factor from rabbit and canine arteries. Circ Res 1986; 58:531-8.
57. Radomski MW, Palmer RMJ, Moncada S. An L-arginine:nitric oxide pathway present in human platelets regulates aggregation. Proc Natl Sci USA 1990; 87:5193-7.
58. Moncada S, Higgs EA. Endogenous nitric oxide: physiology, pathology and clinical relevance. Euro J Clin Invest 1991; 21:361-74.
59. Garthwaite J, Garthwaite G, Palmer RMJ et al. NMDA receptor activation induces nitric oxide synthesis from arginine in rat brain slices. Eur J Pharmacol 1989; 172:413-6.
60. Drapier J, Hibbs J Jr. Murine cytotoxic activated macrophages inhibit aconitase in tumor cells. Inhibition involves the iron-sulfur proesthetic group and is reversible. J Clin Invest 1986; 78:790-7.
61. Lepoivre M, Chenais B, Yapop A et al. Alterations of ribonucleotide reductase activity following induction of the nitrite-generating pathway in adenocarcinoma cells. J Biol Chem 1990; 265:14143-9.
62. Lancaster Jr JR, Hibbs Jr JB. EPR demonstration of iron-nitrosyl complex formation by cytotoxic activated macrophages. Proc Natl Acad Sci USA 1990; 87:1223-7.
63. Pellat C, Henry Y Drapier JC. IFNγ-activated macrophages: detection by electron paramagnetic resonance of complexes between L-arginine-derived nitric oxide and nonheme iron proteins. Biochem Biophys Res Commun 1990; 166:119-25.
64. Hibbs JB Jr, Taintor RR, Vavrin Z. Iron depletion: possible cause of tumor cell cytotoxicity induced by activated macrophages. Biochem Biophys Res Commun 1984; 123:716-23.
65. Beckman JS, Beckman TW, Chem J et al. Apparent hydroxyl radical production by peroxynitrite: implications for endothelial injury from nitric oxide and superoxide. Proc Natl Acad Sci USA 1990; 87:1620-4.
66. Ignarro LJ, Lippton J, Edwards JC et al. Mechanism of vascular smooth muscle relaxation by organic nitrates, nitrites, nitroprusside and nitric oxide: evidence for the involvement of S-nitrosothiols as active intermediates. J Pharmacol Exp Ther 1981; 218:739-49.
67. Keaney JF Jr, Simon DI, Stamler JS et al. NO forms an adduct with serum albumin that has endothelium-derived relaxing factor-like properties. J Clin Invest 1993; 91:1582-9.
68. Gaston B, Reilly J, Drazen JM et al. Endogenous nitrogen oxides and bronchodilator S-nitrosothiols in human airways. Proc Natl Acad Sci USA 1993; 90:10957-61.

69. McDonald B, Reep B, Lapetina EG et al. Glyceraldehyde-3-phosphate dehydrogenase is required for the transport of nitric oxide in platelets. Proc Natl Acad Sci USA 1993; 90:11122-6.
70. Marsden PA, Heng HHQ, Scherer SW et al. Structure and chromosomal localization of the human constitutive endothelial nitric oxide synthase gene. J Biol Chem 1993; 268:1-11.
71. Chartrain NA, Geller DA, Koty PP et al. Human inducible nitric oxide synthase gene. J Biol Chem (in press).
72. Forstermann J, Pollock JS, Schmidt HHHW et al. Calmodulin-dependent endothelium-derived relaxing factor/nitric oxide synthase is present in the particulate and cytosolic fractions of bovine aortic endothelial cells. Proc Natl Acad Sci USA 1991; 88:1788-92.
73. Mishida K, Harrison DG, Navas JP et al. Molecular cloning and characterization of the constitutive bovine aortic endothelial cell nitric oxide synthase. J Clin Invest 1992; 90:2092-6.
74. Magee T, Hanley T. Protein modification. Sticky fingers and CAAX boxes. Nature 1988; 335:114-5.
75. Marsden PA, Schappert KT, Chen HS et al. Molecular cloning and characterization of human endothelial nitric oxide synthase. FEBS Lett 1992; 307:287-93.
76. Michel T, Li GK, Busconi L. Phosphorylation and subcellular translocation of endothelial nitric oxide synthase. Proc Natl Acad Sci USA 1993; 90:6252-6.
77. Geller DA, Nussler AK, Di Silvio M et al. Cytokines, endotoxin, and glucocorticoids regulate the expression of inducible nitric oxide synthase in hepatocytes. Proc Natl Acad Sci USA 1993; 90:522-6.
78. Stuehr DJ, Marletta MA. Mammalian nitrate biosynthesis: mouse macrophages produce nitrite and nitrate in response to *Escherichia coli* lipopolysaccharide. Proc Natl Acad Sci USA 1985; 82:7738-42.
79. Stuehr DJ, Marletta MA. Induction of nitrite/nitrate synthesis in murine macrophages by BCG infection, lymphokines or interferon-γ. J Immunol 1987; 139:518-25.
80. Busse R, Mulsch A. Induction of nitric oxide synthase by cytokines in vascular smooth muscle cells. FEB Lett 1990; 275:87-90.
81. Curran RD, Billiar TR, Stuehr DJ et al. Hepatocytes produce nitrogen oxides from L-arginine in response to inflammatory products of Kupffer cells. J Exp Med 1989; 170:1769-74.
82. Werner-Felmayer G, Werner ER, Fuchs D et al. Tetrahydrobiopterin-dependent formation of nitrite and nitrate by murine fibroblasts. J Exp Med 1990; 172:1599-1607.
83. Gross SS, Jaffe EA, Levi R et al. Cytokine-activated endothelial cells express an isotype of nitric oxide synthase which is tetrahydrobiopterin-dependent, calmodulin-independent and inhibited by arginine analogs with a rank-order of potency characteristic of activated macrophages. Biochem Biophys Res Commun 1991; 178:823-9.
84. Schmidt K, Werner ER, Mayer B et al. Tetrahydrobiopterin-dependent formation of endothelium-derived relaxing factor (nitric oxide) in aortic endothelial cells. Biochem J 1992; 281:297-300.
85. Werner-Felmayer G, Werner ER, Fichs D et al. Pteridine biosynthesis in human endothelial cells: impact on nitric oxide-mediated formation of cyclic GMP. J Biol Chem 1993; 268:1842-6.
86. Werner ER, Werner-Felmayer G, Wachter H. Tetrahydrobiopterin and cytokines. PSEBM 1993; 203:1-12.
87. Kwang JB, Thiel BA, Lucas S et al. Macrophage nitric oxide synthase subunits: purification, characterization, and role of prosthetic groups and substrate in modulating their association into a dimeric enzyme. J Biol Chem 1993; 28: 21120-9.
88. Closs EI, Lyons CR, Mitchell M et al. Expression of cationic amino acid transporters in NO-producing macrophages. Endothelium 1993; 1:s16(abstract).
89. Nussler AK, Billiar TR, Liu ZZ et al. Coinduction of nitric oxide synthase and argininosuccinate synthetase in a murine macrophage cell line. Implications for regulation of nitric oxide production. J Biol Chem 1994; in press.
90. Cho HJ, Xie Q, Calaycay J et al. Calmodulin is a subunit of nitric oxide synthase from macrophages. J Exp Med 1992; 176: 599-604.

91. Geller DA, Lowenstein CJ, Shapiro RA et al. Molecular cloning and expression of inducible nitric oxide synthase from human hepatocytes. Proc Natl Acad Sci USA 1993; 90:3491-5.
92. Radomski MW, Palmer RMJ, Moncada S. Glucocorticoids inhibit the expression of an inducible, but not the constitutive, nitric oxide synthase in vascular endothelial cells. Proc Natl Acad Sci USA 1990; 87:10043-7.
93. Di Rosa M, Radomski M, Carnuccio R et al. Glucocorticoids inhibit the induction of nitric oxide synthase in macrophages. Biochem Biophys Res Commun 1990; 172: 1246-52.
94. Nussler AK, Billiar TR. Inflammation, immunoregulation, and inducible nitric oxide synthase. J Leuk Biol 1993; 54:171-8.
95. Xie QW, Cho HJ, Calaycay J et al. Cloning and characterization of inducible nitric oxide synthase from mouse macrophages. Science 1992; 256:225-8.
96. Lowenstein CJ, Glatt CS, Bredt DS et al. Cloned and expressed macrophage nitric oxide synthase contrasts with the brain enzyme. Proc Natl Acad Sci USA 1992; 89:6711-5.
97. Lyons CR, Orloffs GJ, Cunningham JM. Molecular cloning and functional expression of an inducible nitric oxide synthase from a murine macrophage cell line. J Biol Chem 1992; 267:6370-4.
98. Denis M. Tumor necrosis factor and granulocyte macrophage-colony stimulating factor stimulate human macrophages to restrict growth of virulent *Mycobacterium avium* and to kill avirulent *M. avium*: killing effector mechanism depends on the generation of reactive nitrogen intermediates. J Leuk Biol 1991; 49:380-7.
99. Munoz-Fernandez MA, Fernandez MA, Fresno M. Activation of human macrophages for the killing of intracellular Trypa–nosoma cruzi by TNF-α and IFN-γ through a nitric oxide-dependent mechanism. Immunol Lett 1992; 33:35-40.
100. Hunt NCA, Goldin RD. Nitric oxidc production by monocytes in alcoholic liver disease. J Hepatol 1992; 14:146-50.
101. Mellouk S, Green SJ, Nacy CA et al. IFN-γ inhibits development of *Plasmodium berghei* exoerythrocytic stages in hepatocytes by an L-arginine-dependnent effector mechanism. J Immunol 1991; 146:3971-6.
102. Nussler AK, Drapier JC, Renia L et al. L-arginine-dependent destruction of intrahepatic malaria parasites in response to tumor necrosis factor and/or interleukin-6 stimulation. Eur J Immunol 1991; 21:227-30.
103. Nussler AK, Renia L, Pasquetto V et al. In vivo induction of the nitric oxide pathway in hepatocytes after injection with irradiated malaria sporozoites, malaria blood parasites or adjuvants. Eur J Immunol 1993; 23:882-7.
104. Murad F, Arnold WP, Mittal CK et al. Properties and regulation of guanylate cyclase and some proposed functions for cyclic GMP. Adv Cyclic Nucleotide Res 1979; 11:175-204.
105. Kukovetz WR, Poch G, Holzmann S. Cyclic nucleotides and relaxation of vascular smooth muscle. In: Vanhoutte PM, Leusen I, eds. Vasodilatation. New York: Raven, 1981:339-53.
106. Ignarro LJ, Gruetter CA, Hyman AL et al. Molecular mechanisms of vasodilatation. In: Poste G, Crooke ST, eds. Dopamine receptor agonists. New York: Plenum, 1984: 259-88.
107. Ignarro LJ, Lippton J, Edwards JC et al. Mechanism of vascular smooth muscle relaxation by organic nitrates, nitrites, nitroprusside and nitric oxide: evidence for the involvement of S-nitrosothiols as active intermediates. J Pharmacol Exp Ther 1981; 218:739-49.
108. Gruetter CA, Barry BK, McNamara DB et al. Relaxation of bovine coronary artery and activation of coronary arterial guanylate cyclase by nitric oxide, nitroprusside and a carcinogenic nitrosoamine. J Cyclic Nucleotide Res 1979; 5:211-24.
109. Gruetter CA, Gruetter DY, Lyon JE et al. Relationship between cyclic guanosine 3':5'-monophosphate formation and relaxation of coronary arterial smooth muscle by glyceryl trinitrate, nitroprusside, nitrite and nitric oxide: effects of methylene blue and methemoglobin. J Pharmacol Exp Ther 1981; 219:181-6.
110. Mellion BT, Ignarro LJ, Ohlstein EH et al.

Evidence for the inhibitory role of guanosine 3',5'-monophosphate in ADP-induced human platelet aggregation in the presence of nitric oxide and related vasodilators. Blood 1981; 57:946-55.
111. Mellion BT, Ignarro LJ, Myers CB et al. Inhibition of human platelet aggregation by S-nitrosothiols. Heme-dependent activation of soluble guanylate cyclase and stimulation of cyclic GMP accumulation. Mol Pharmacol 1983; 23:653-64.
112. Ignarro LJ. Endothelium-derived nitric oxide: actions and properties. FASEB 1989; 3:31-6.
113. Martin W, Villani GM, Jothianandan E et al. Selective blockade of endothelium-dependent and glyceryl trinitrate-induced relaxation by hemoglobin and by methylene blue in the rabbit aorta. J Pharmacol Exp Ther 1985; 232:708-16.
114. Lancaster JR Jr. Nitric oxide in cells. Amer Scientist 1992; 80:248-59.
115. Aisaka K, Gross SS, Griffith OW et al. N^G-methylarginine, an inhibitor of endothelium-derived nitric oxide synthesis, is a potent pressor agent in the guinea pig: does nitric oxide regulate blood pressure in vivo? Biochem Biophys Res Commun 1989; 160:881-6.
116. Rees DD, Palmer RMJ, Moncada S. Role of endothelium-derived nitric oxide in the regulation of blood pressure. Proc Natl Acad Sci USA 1989; 86:3375-8.
117. Pohl U, Busse R, Kuon E et al. Pulsatile perfusion stimulates the release of endothelial autacoids. J Appl Cardiol 1986; 1:215-35.
118. Rubanyi GM, Romero JC, Vanhoutte PM. Flow-induced release of endothelium-derived relaxing factor. Am J Physiol 1986; 250:H1145-9.
119. Linder L, Kiowski W, Buhler RF et al. Indirect evidence for release of endothelium-derived relaxing factor in human forearm circulation in vivo. Blunted response in essential hypertension. Circulation 1990; 81:1762-7.
120. Panza JA, Quyyumi AA, Brush Jr JE, et al. Abnormal endothelium-dependent vascular relaxation in patients with essential hypertension. N Engl J Med 1990; 323:22-7.
121. Whittle BJR, Lopez-Belmonte J, Rees DD. Modulation of the vasopressor actions of acetylcholine, bradykinin, substance P and endothelin in the rat by a specific inhibitor of nitric oxide formation. Br J Pharmacol 1989; 98:646-52.
122. Vallance P, Collier J, Moncada S. Effects of endothelium-derived nitric oxide on peripheral arteriolar tone in man. Lancet 1989; 2:997-1000.
123. Gardiner SM, Compton AM, Bennett T et al. Control of regional blood flow by endothelium-derived nitric oxide. Hypertension 1990; 15:486-92.
124. Gardiner SM, Compton AM, Bennett T et al. Persistent haemodynamic changes following prolonged infusions of N^G-monomethyl-L-arginine (L-NMMA) in conscious rats. In : Moncada S Higgs EA, eds. Nitric Oxide from L-arginine: A Bioregulatory System. Amsterdam: Elsevier, 1990:489-91.
125. Ware JA, Johnson PC, Smith M et al. The effect of common agonists on cytoplasmic ionized calcium concentration in platelets: measurements with quin2 and aequorin. J Clin Invest 1986; 77:878-86.
126. Corbett JA, Tilton RG, Chang K et al. Aminoguanidine, a novel inhibitor of nitric oxide formation, prevents diabetic vascular dysfunction. Diabetes 1992; 41:552-6.
127. Mulligan MS, Hevel JM, Marletta et al. Tissue injury caused by deposition of immune complexes is L-arginine dependent. Proc Natl Acad Sci USA 1991; 88:6338-42.
128. Durante W, Sen AK, Sunahara FA. Impairment of endothelium-dependent relaxation in aortae from spontaneously diabetic rats. Br J Pharmacol 1988; 94:463-8.
129. Kappagoda T, Jayakody L, Rajotte R et al. Endothelium-dependent relaxation to acetylcholine in the aorta of streptozocin induced diabetic rat and the BB-diabetic rat. Clin Invest Med 1989;12:187-93.
130. Kamata K, Miyata N, Kasuya Y. Impairment of endothelium-dependent relaxation and changes in levels of cyclic GMP in aorta from streptozotocin-induced diabetic rats. Br J Pharmacol 1989; 97:614-8.
131. Oyama Y, Kawasaki H, Hattori Y et al. Attenuation of endothelium-dependent relaxation in aorta from diabetic rats. Eur J Pharmacol 1986; 131:75-8.

132. Meraji S, Jayakody L, Senaratne MPJ et al. Endothelium-dependent relaxation in aorta of BB rat. Diabetes 1987; 36:978-81.
133. de Tejada IS, Goldstein I, Azadzoi K et al. Impaired neurogenic and endothelium-mediated relaxation of penile smooth muscle in diabetic men with impotence. N Engl J Med 1989; 320:1025-30.
134. Brink C, Duncan PG, Douglas JS. Decreased vascular sensitivity to histamine during aging. Agents Actions 1984; 14:8-10.
135. Gascho JA, Ganelli C, Zelis R. Aging reduces venous distensibility and the venodilatory response to nitroglycerin in normal subjects. Am J Cardiol 1989; 63:1267-70.
136. Bucala R, Tracey KJ, Cerami A. Advanced glycosylation products quench nitric oxide and mediate defective endothelium-dependent vasodilatation in experimental diabetes. J Clin Invest 1991; 87:432-8.
137. Hogan M, Cerami A, Bucala R. Advanced glycosylation endproducts block the antiproliferative effect of nitric oxide. Role in the vascular and renal complications of diabetes mellitus. J Clin Invest 1992; 90: 1110-5.
138. Chester AH, O'Neil GS, Moncada S et al. Low basal and stimulated release of nitric oxide in atherosclerotic epicardial coronary arteries. Lancet 1990; 336:897-900.
139. Verbeuren TJ, Jordaens FH, Zonnekeyn LL et al. Effect of hypercholesterolemia on vascular reactivity in the rabbit. Circ Res 1986; 58:552-64.
140. Verbeuren TJ, Jordaens FH, Van Hove CE et al. Release and vascular activity of endothelium-derived relaxing factor in atherosclerotic rabbit aorta. Eur J Pharmacol 1990; 191:173-84.
141. Drexler H, Zeiher AM, Wollshlager H et al. Flow-dependent coronary artery dilatation in humans. Circulation 1989; 80: 466-74.
142. Dinh-Xuan AT, Higenbottam TW, Pepke-Zaba J et al. Reduced endothelium-dependent relaxation of cystic fibrosis pulmonary arteries. Eur J Pharmacol 1989; 163:401-3.
143. Dinh-Xuan AT, Higenbottam TW, Pepke-Zaba J et al. Physiological and histological correlates of pulmonary endothelium-dependent relaxation in human chronic obstructive lung disease. Thorax 1990; 45:810.
144. Dinh-Xuan AT, Higenbottam TW, Clelland CA et al. Impairment of pulmonary endothelium-dependent relaxation in patients with Eisenmenger's syndrome. Br J Pharmacol 1990; 90:9-10.
145. Houston DS, Shepherd JT, Vanhoutte PM. Adenine nucleotides, serotonin, and endothelium-dependent relaxations to platelets. Am J Physiol 1985; 248:H389-95.
146. Ku DD. Coronary vascular reactivity after acute myocardial ischemia. Science 1982; 218:576-8.
147. Van Benthuysen KM, McMurtry IF, Horowitz LD. Reperfusion after acute coronary occlusion in dogs impairs endothelium-dependent relaxation to acetylcholine and augments contractile reactivity in vitro. J Clin Invest 1987; 79:265-74.
148. Mehta JL, Nichols WW, Donnelly WH et al. Impaired canine coronary vasodilator response to acetylcholine and bradykinin after occlusion-reperfusion. Circ Res 1989; 64:43-54.
149. Mehta JL, Nichols WW, Donnelly WH et al. Protection by superoxide dismutase from myocardial dysfunction and attenuation of vasodilatory reserve after coronary occlusion and reperfusion in dog. Circ Res 1989; 65: 1283-95.
150. Olafsson B, Forman MB, Puett DW et al. Reduction of reperfusion injury in the canine preparation by intracoronary adenosine: importance of the endothelium and the no-reflow phenomenon. Circulation 1987; 76:1135-45.
151. Mehta JL, Lawson DL, Nichols WW. Attenuated coronary relaxation after reperfusion: effects of superoxide dismutase and TxA_2 inhibitor U-63557. Am J Physiol 1989; 257:H1240-6.
152. Tsao PS, Aoki N, Lefer DJ et al. Time course of endothelial dysfunction and myocardial injury during myocardial ischemia and reperfusion in the cat. Circulation 1990; 82:1402-12.
153. Lefer AM, Tsao PS, Lefer DJ et al. Role of endothelial dysfunction in the pathogenesis of reperfusion injury after myocardial ischemia. FASEB J 1991; 5:2029-34.
154. Rubanyi GM, Vanhoutte PM. Superoxide anions and hyperoxia inactivate endothe-

lium-derived relaxing factor. Am J Physiol 1986; 250:H822-456.
155. Moncada S, Palmer RMJ, Gryglewski RJ. Mechanism of action of some inhibitors of endothelium-derived relaxing factor. Proc Natl Acad Sci USA 1986; 83:9164-8.
156. Mugge A, Elwell JH, Peterson TE et al. Release of intact endothelium-derived relaxing factors depends on endothelial superoxide dismutase activity. Am J Physiol 1991; 260:C219-25.
157. Gorren ACF, de Boer E, Wever R. The reaction of nitric oxide with copper proteins and the photodissociation of copper-NO complexes. Biochem Biophys Acta 1987; 916:38-47.
158. Freeman B, Crapo JD. Biology of disease-free radicals and tissue injury. Lab Invest 1982; 47:412-26.
159. Flaherty JT, Weisfeldt ML. Reperfusion injury: free radical. Biol Med 1988; 3:409-19.
160. McCord JM. Oxygen-derived free radicals in postischemic tissue injury. N Engl J Med 1985; 312:159-63.
161. Zweier JL, Rayburn BK, Flaherty JT et al. Recombinant superoxide dismutase reduces oxygen free radical concentrations in reperfused myocardium. J Clin Invest 1987; 80:1728-34.
162. Zweier JL, Flaherty JT, Weisfeldt ML. Observation of free radical generation in the postischemic heart. Proc Natl Acad Sci USA 1987; 84:1404-7.
163. Braunwald E, Kloner RA. Myocardial reperfusion: a double-edged sword? J Clin Invest 1985; 76:1713-9.
164. Siegfried MR, Erhardt J, Rider T et al. Cardioprotection and attenuation of endothelial dysfunction by organic nitric oxide donors in myocardial ischemia-reperfusion. J Pharmacol Exp Ther 1991; 260:668-75.
165. Vallance P, Collier J, Moncada S. Nitric oxide synthesized from L-arginine mediates endothelium-dependent dilatation in human veins. Cardiovasc Res 1989; 23:1053-7.
166. Matheis G, Sherman MP, Buckberg GD et al. Role of L-arginine-nitric oxide pathway in myocardial reoxygenation injury. Am J Physiol 1992; 262:H616-20.
167. McCord JM, Fridovich I. The reduction of cytochrome C by milk xanthine oxidase. J Biol Chem 1968; 243:5753-60.
168. McCord JM. Oxygen-derived free radicals in postischemic tissue injury. N Engl J Med 1985; 312:159-63.
169. Roy RS, McCord JM. Superoxide and ischemia: conversion of xanthine dehydrogenase to xanthine oxidase. In: Greenwald R, Cohen G, eds. Oxy radicals and their scavenger systems. Vol. 2. Cellular and molecular aspects. New York: Elsevier Science, 1983:145-53.
170. Smith JK, Carden DL, Grisham MB et al. Role of iron in postischemic microvascular injury. Am J Physiol 1989; 256:H1472-77.
171. Zweier JL, Flaherty JT, Weisfeldt ML. Direct measurement of free radical generation following reperfusion of ischemic myocardium. Proc Natl Acad Sci USA 1987; 84:1404-7.
172. Rubanyi GM, Ho EH, Cantor EH et al. Cytoprotective function of nitric oxide: activation of superoxide radicals produced by human leukocytes. Biochem Biophys Res Commun 1991; 181:1392-7.
173. Wink DA, Hanbauer I, Krishna MC et al. Nitric oxide protects against cellular damage and cytotoxicity from reactive oxygen species. Proc Natl Acad Sci USA 1993; 90:9813-7.
174. Hearse DJ, Humphrey ST, Bullock GR. The oxygen paradox and the calcium paradox: two facets of the same problem? J Mol Cell Cardiol 1976; 10:641-8.
175. Clancy RM, Leszczynska-Piziak J, Abramson SB. Nitric oxide, an endothelial cell relaxation factor, inhibits neutrophil superoxide anion production via a direct action on the NADPH oxidase. J Clin Invest 1992; 90:1116-21.
176. Johnson G, Tsao PS, Lefer AM. Cardioprotective effects of authentic nitric oxide in myocardial ischemia with reperfusion. Critical Care Med 1991; 19:244-52.
177. Billiar TR, Curran RD, Stuehr DJ et al. Inducible cytosolic enzyme activity for the production of nitrogen oxides from L-arginine in hepatocytes. Biochem Biophys Res Commun 1990; 168:1034-40.
178. Curran RD, Billiar TR, Stuehr DJ et al. Multiple cytokines are required to induce hepatocyte nitric oxide production and inhibit total protein synthesis. Ann Surg 1990; 212:462-71.

179. Billiar TR, Curran RD, Harbrecht BG et al. Modulation of nitrogen oxide synthesis in vivo: N^G-monomethyl-L-arginine inhibits endotoxin-induced nitrite/nitrate biosynthesis while promoting hepatic damage. J Leuk Biol 1990; 48:565-9.
180. Arthur MJP, Bentley IS, Tanner AR et al. Oxygen-derived free radicals promote hepatic injury in the rat. Gastroenterology 1985; 89:1114-22.
181. Harbrecht BG, Stadler J, Demetris AJ et al. Nitric oxide and prostacyclin interact to prevent hepatic damage during murine endotoxemia. Am J Physiol 1994; in press.
182. Saran M, Michel C, Bors W. Reaction of NO with $O_2^{\cdot-}$: implications for the actions of endothelium-derived relaxing factor (EDRF). Free Radicals Res Commun 1990; 10:221-6.
183. Moncada S, Palmer RMJ, Higgs EA. The discovery of nitric oxide as the endogenous nitrovasodilatory. Hypertension 1988; 12:365-72.
184. Rosen GM, Freeman BA. Detection of superoxide generated by endothelial cells. Proc Natl Acad Sci USA 1984; 81:7269-73.
185. Radi R, Beckman JS, Bush KM et al. Peroxynitrite oxidation of sulfhydryls: the cytotoxic potential of superoxide and nitric oxide. J Biol Chem 1991; 266:4244-50.
186. Radi R, Beckman JS, Bush KM et al. Peroxynitrite-induced membrane lipid peroxidation: the cytotoxic potential of superoxide and nitric oxide. Arch Biochem Biophys 1991; 288:481-7.
187. Beckman JS, Ischiropoulos H, Zhu H et al. Kinetics of superoxide dismutase- and iron-catalyzed nitration of phenolics by peroxynitrite. Arch Biochem Biophys 1992; 298:438-45.
188. Ischiropoulos H, Zhu L, Chen J et al. Peroxynitrite-mediated tyrosine nitration catalyzed by superoxide dismutase. Arch Biochem Biophys 1992; 298:431-7.
189. Zhu L, Gunn C, Beckman JS. Bactericidal activity of peroxynitrite. Arch Biochem Biophys 1992; 298:452-7.
190. Prutz WA, Monig H, Butler J et al. Reactions of nitrogen dioxide in aqueous model systems: oxidation of tyrosine units in peptides and proteins. Arch Biochem Biophys 1985; 243:125-34.
191. Burkitt MJ, Gilbert BC. Model studies of the iron-catalysed Haber-Weiss cycle and the ascorbate-driven Fenton reaction. Free Radical Res Commun 1990; 10:265-80.

CHAPTER 8

Iron Delocalization and Iron Binding

Bo E. Hedlund

This chapter will be divided into two main sections. The first will deal with the concept of iron delocalization and highlight selected experimental evidence that supports the contention that reactive iron is not only released during ischemia and reoxygenation, but, more importantly, that this mobilized iron plays a major role in the genesis of reperfusion injury. The second section will deal with how the consequences of iron delocalization can be minimized, either pharmacologically or by altering procedural aspects of surgical intervention, to minimize vascular and parenchymal cell injury.

IRON DELOCALIZATION AND REOXYGENATION INJURY

WHAT IS "DELOCALIZED IRON" AND WHY IS IT IMPORTANT?

Throughout evolution, complex organisms have had to deal with two issues relating to iron: utilization and storage. Iron has been incorporated into the most basic functions of life from an early timepoint in evolution, being utilized as a critical component in aerobic metabolism and DNA synthesis. In mammals, iron is either utilized in iron-containing proteins or set aside in iron storage proteins. In addition, a small fraction of the total body iron is normally "in transit" between these two compartments in the form of the transport protein, transferrin. The major compartments of "used iron" are the two hemeproteins, hemoglobin and myoglobin, accounting for about 75–80% of total iron in man. Most of the remaining 20–25% is divided between the two storage proteins, ferritin and hemosiderin. Only about 0.1% of the total body iron is normally in circulation, bound to transferrin. As a consequence of the size of the stored iron compartment in relation to the total binding capacity, it follows that "mobilization" of a extremely small fraction of ferritin-derived iron is sufficient to locally, or in

Ischemia-Reperfusion Injury of Skeletal Muscle, edited by Gary A. Fantini, M.D.;

extreme cases, systemically exceed transferrin binding capacity.

The term "iron delocalization" refers to inappropriate release of iron from either of the two major iron pools, leaving a transient pool of free iron either in the intra- or extracellular compartment. Excessive release into the vascular compartment can locally or systemically saturate the transferrin binding capacity. It is implied in this definition that this free iron is reactive in that it participates in a variety of oxygen and lipid mediated reactions, usually acting in a catalytic fashion. Among the most important and potentially damaging biological reactions catalyzed by this transition metal results in the production of hydroxyl radical [$OH^{\bullet}$], a potent oxidant, from superoxide generating systems and is described by the Haber-Weiss equation:[1]

$$O_2^{-\bullet} + H_2O_2 \longrightarrow OH^{\bullet} + OH^- + O_2$$

In the absence of a catalyst, however, this reaction would proceed too slowly to be of biological significance. In the presence of certain transition metals (M), hydroxyl radicals may be generated more rapidly through Fenton chemistry[2]

$$Mn^+ + H_2O_2 \longrightarrow M^{(n+1)+} + OH^{\bullet} + OH^-$$

It is now recognized that, under physiologic conditions, iron (Fe) is the most important metal catalyst in this regard,[3,4] and the iron-catalyzed Haber-Weiss reaction is represented by the following sequence of reactions:

$$Fe^{3+} + O_2^{-\bullet} \longrightarrow Fe^{2+} + O_2$$

$$2O_2^{-\bullet} + 2H^+ \longrightarrow H_2O_2 + O_2$$

$$Fe^{2+} + H_2O_2 \longrightarrow Fe^{3+} + OH\bullet + OH^-$$

In addition to its role in the generation of hydroxyl radicals, iron may also form reactive molecular complexes with oxygen, ferryl ($Fe^{2+}O$) and perferryl ($Fe^{2+}O_2$) species, and may also directly catalyze the decomposition of lipid hydroperoxides to alkoxy ($LO\bullet$) and peroxy ($LOO\bullet$) radicals:[5,6]

$$2LOOH \xrightarrow{\text{iron}} LO\bullet + LOO\bullet + H_2O$$

$$LH + LOO^{\bullet} \longrightarrow L^{\bullet} + LOOH$$

$$L^{\bullet} + O_2 \longrightarrow LOO^{\bullet}$$

Therefore, it is evident that this type of "oxidant prone" iron can, by virtue of its catalytic properties, cause considerable tissue injury when present at very low concentration in the postischemic environment.

The two primary sources of reducing equivalents, such as superoxide anion, are activated neutrophils and the enzyme xanthine oxidase, in the presence of a suitable substrate, such as hypoxanthine. This substrate is generated in significant quantities during an ischemic insult when ATP is utilized. A schematic drawing illustrating these events is provided in Figure 8.1. It would seem clear that steps can be taken to reduce postischemic injury caused by catalytically active iron by employing therapeutic interventions which will either neutralize or minimize its release into the tissue.

"In vivo" Sources of Oxidant Prone Iron

As noted above, free iron is highly reactive and therefore, under normal circumstances, iron in the vascular compartment is tightly bound to transferrin. Studies aimed at detecting iron in plasma, using the bleomycin method,[7,8] fail to detect reactive iron in normal plasma. In the intracellular compartment, the situation is less well defined. As much as 5% of the total intracellular iron pool may be free, or "chelatable". A small fraction of ferritin bound iron exists in some form of equilibrium with the cell's chelatable iron pool. This chelatable iron, or the "labile iron pool", which normally is innocuous, may represent the source of iron that leads to oxidant injury in the context of ischemia and reperfusion. Thus, parenchymal cells may contain significant quantities of "loosely" bound iron, which under certain conditions can leak or diffuse into the vascular compartment.

The iron storage protein ferritin is present in most human cells, contains large quantities of iron, and appears to be the primary source of iron released in situations involving ischemia. Several investigators have documented the release of iron from ferritin in "in vitro" systems. An early study by Williams et al[9] demonstrating that iron release from ferritin in the presence of xanthine oxidase and xanthine could be inhibited by superoxide dismutase (SOD) may have been the first study clearly implicating the superoxide anion as the reducing agent in this reaction. Of considerable physiological relevance is the study by Biemond et al[10] in which these workers observed that the activated neutrophil could generate iron from ferritin preparations and that this release was also inhibitable by SOD. Another relevant report in the context was provided by Funk et al,[11] who demonstrated that reduced flavin nucleotides can reduce large quantities of iron from ferritin and that the magnitude of release increases with decreasing pH. In most biochemical studies, ferritin derived from horse spleen was used. Biochemical and biophysical properties of such "purified" ferritin preparations may differ from the "undisturbed" intracellular ferritin, and extrapolations from results obtained in buffered solutions to potential similar reactions occurring "in vivo" should not be done without keeping this limitation in mind. The topic of ferritin as source for iron leading to oxidative damage was recently examined in a comprehensive review by Reif.[12]

Another potential source of oxidant prone iron, although less likely to be involved in most situations associated with reperfusion injury, is heme bound iron. There is some controversy whether or not heme bound iron can act as an oxidant. There is considerable evidence for the formation of higher valence states of iron when in the presence of activated

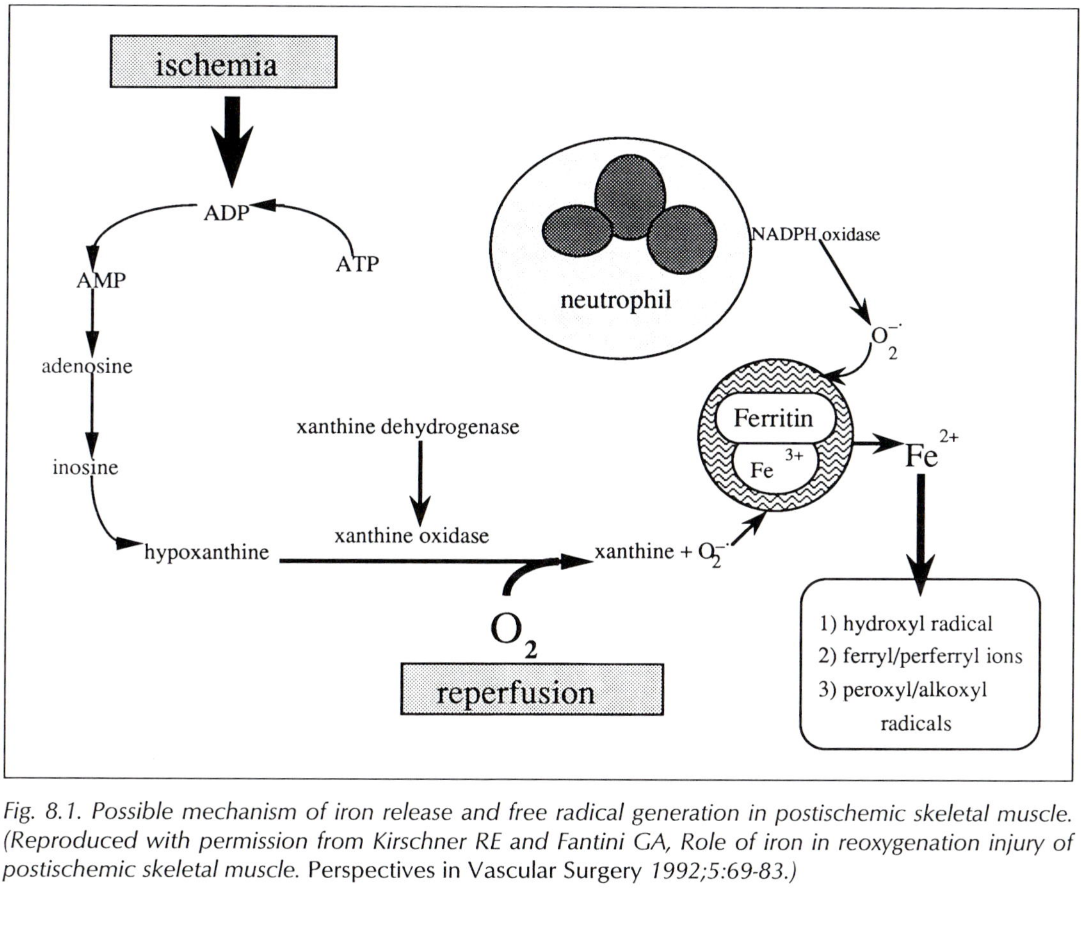

Fig. 8.1. Possible mechanism of iron release and free radical generation in postischemic skeletal muscle. (Reproduced with permission from Kirschner RE and Fantini GA, Role of iron in reoxygenation injury of postischemic skeletal muscle. Perspectives in Vascular Surgery *1992;5:69-83.)*

oxygen species, such as ferryl and perferryl complexes. The molecular combinations of iron and oxygen are undoubtedly chemically reactive. However, the degree to which these complexes are important "in vivo" oxidants is uncertain.[13] Experimentally one can, in principle, determine whether or not free or heme bound iron is responsible for the oxidant event by using an iron specific chelator, such as deferoxamine (DFO). This specific chelator would only bind to free iron and should have little effect on chemical reactions utilizing perferryl complexes. When hemeproteins are extravasated or bound to membranes, e.g., as hemichrome, their potential for oxidant damage increases dramatically.[14] The heme ring is in itself easily broken by radical reactions occurring at or adjacent to the bound iron, an event which would lead to release of iron. It is conceivable that this type of "iron leakage" from hemoglobin is responsible for the neurotoxicity of hemoglobin.[15]

Iron Equilibria in Cultured Cells

Cells growing in culture represent an experimental system stripped of much of the complexity of experiments using whole animals, or for that matter, an isolated organ. Two examples of cell culture studies relating to oxidant injury have been selected for discussion. In one, endothelial cells growing in culture are used as target cells for activated neutrophils, and, in the other, epithelial cells in culture are exposed to an hypoxic insult, followed by reoxygenation, thus mimicking ischemia, or, hypoxia followed by reperfusion. Both these studies also focus on iron as a component in the oxidant injury and demonstrate that valuable information pertaining to the mechanism of such injury can be derived from cell culture studies.

In an attempt to answer the question of the role of iron in protection of endothelial cells in culture against neutropohil mediated killing, Gannon et al[16] preincubated these cells with DFO before exposure to activated neutrophils. It was noted that the DFO-exposed endothelial cells became resistant to neutrophil mediated killing. Furthermore, they noted that the oxidative burst of the neutrophil was unaffected by the presence of DFO, indicating that the DFO-exposed endothelial cells have become resistant to the otherwise lethal oxidative burst. The authors also demonstrated uptake of minute quantities of DFO, which led them to conclude that the protective effect of DFO is exerted within the cytoplasmic domain. The physiological ramifications of the study by Gannon et al are somewhat uncertain. Do the results indicate that the protective effect of an antioxidant is limited to those agents that penetrate the cell membrane? It is impossible to achieve concentrations in vivo approaching those in used in this study. As will be discussed in subsequent sections, results from many in vivo experiments indicate that membrane permeability is not a prerequisite of protection against iron mediated oxidant injury.

More recently Paller and Hedlund[17] carried out a study in which the effect of a hypoxic insult on cells in culture was investigated. These investigators utilized rat kidney proximal epithelial cells that were exposed to 60 minutes of hypoxia (H) followed 30 minutes of reoxygenation (R). The experiment represents a cell culture version of a typical ischemia/reperfusion study of an isolated organ. Release of iron was quantitated. Cell viability was assessed by release of lactate dehydrogenase. Protection to H/R induced injury was assessed using either free DFO or a high molecular weight form, HES-DFO, where the drug is covalently attached to hydroxyethyl starch of an average molecular weight of 80,000 Dalton. The authors noted a roughly 5-fold increase of iron release into the medium from cells exposed to H/R compared to cells incubated under normal conditions; and, when iron was measured in cell homogenates, there was a 60-70% increase in iron measurable with ferrozine in cells that had been exposed to H/R compared to control cells. Finally, both DFO and HES-DFO provided nearly equal protection against cell death caused by hypoxia/reperfusion. In summary, these cells, when exposed to an hypoxic insult, release iron into the supernatant, and more "chelatable" iron can be measured in cell homogenates. Also, both DFO, a small chelator with molecular weight 560 Dalton, and a polymeric form of this chelator, afford similar protection against H/R injury. These two cell culture experiments provide evidence that questions of iron release and cellular injury can be experimentally approached and provide valuable information. However, extrapolation of these results to "in vivo" conditions remains risky.

Iron, Ischemia and Reperfusion in Isolated Organs

The isolated, buffer perfused, organ represents the next step up in complexity from cells in culture. Of this type of experimental system, the Langendorff preparation of the rat heart is one of the most widely used and represents the system in which the reperfusion injury was first clearly demonstrated by Hearse.[18] Among the multitude of studies of this type dealing with oxidant injury and its prevention, studies from two laboratories have been chosen for a more detailed discussion.

Studies using the Langendorff preparation tend to focus on functional impairment of the myocardium caused by some external insult. Changes in myocardial integrity are assessed by the ability of the isolated heart to carry out its normal workload, and this is typically done using a small balloon inserted into the atrium to measure volume/pressure parameters of the beating heart. For example, Tsao and Lefer[19] focused on alterations in endothelial function following exposure of isolated hearts to varying degrees of ischemic and hypoxic insults, rather than the normal myocardial workload endpoint. Although this study does not discuss iron, it is important for its focus on endothelial cell changes and the study provides clear evidence for the profound difference between ischemic and anoxic insults. Tsao and Lefer measured changes in responsivity to the endothelium dependent vasoconstrictor, acetylcholine, by continuous measurement of coronary perfusion pressure using a transducer. They noted that an ischemic insult as short as 2.5 minutes at 37°C was sufficient to significantly impair endothelial function. This observation contrasted with the lack of effect, for up to 20 minutes, when a nonendothelium-dependent agent, nitroglycerin, was used. It should be noted that the endothelial dysfunction was significantly attenuated when superoxide dismutase was added to the buffer within one minute following initiation of ischemia. In a second set of experiments, the investigators used hypoxic perfusion as opposed to ischemic "no flow" to achieve an anti-oxidant insult. The hearts were perfused with buffer equilibrated with 95% nitrogen and 5% CO_2, which lowered the coronary perfusate pO_2 by about 80%. Under these conditions, significant impairment of endothelial function also occurred, but the magnitude of the effect was less than during ischemia. Thus, 20 minutes of hypoxic/anoxic perfusion caused as much loss in endothelial function as 2.5 minutes of ischemia. Due to experimental difficulties of achieving total anoxia, it is difficult to determine whether or not totally anoxic perfusion would lead to more pronounced injury. However, it is strongly suggested by this study that a "no flow", i.e. a true ischemic insult, is more deleterious than severely hypoxic perfusion.

A more recent study by Voogd et al[20] from Koster's laboratory in Rotterdam is particularly important and informative both in terms of iron delocalization in the context of ischemia and the differential injury caused by ischemia and hypoxic perfusion. The authors quantitated the magnitude of the "low molecular weight iron pool" in the perfused rat heart, and have clearly demonstrated the release of substantial quantities of chelatable iron from the buffer perfused myocardium after an ischemic insult. Voogd et al also compared the magnitude of iron release caused by two different types of hypoxic or anoxic challenge. Unlike Lefer and Tsao,[15] Voogd and collaborators focused entirely on functional, i.e., contractile, integrity of their isolated heart preparations. A more severe ischemic insult was required to cause significant impairment of the ability of the isolated heart to perform work than was the case when endothelial integrity was assessed. In one set of experiments, Voogd and coworkers exposed hearts to total, no-flow ischemia, and in another set of experiments the hearts were perfused with oxygen free buffer. They noted an increase from 2 to 54 nmoles of chelatable low molecular weight iron per gram of wet weight in hearts exposed to 45 minutes of warm ischemia. Even at 15 minutes of warm ischemia, as much as 15 nmoles was observed. Only about 7 nmoles per gram of wet weight of iron is released from hearts exposed to 45 minutes of anoxic perfusion when these hearts were reperfused with oxygenated buffer. In terms of functional damage, the same group had, in an earlier publication,[21] determined that hearts exposed to a 15 minute ischemic insult only recover 20% of their contractility following reperfusion, while hearts exposed to 45 minutes of anoxic perfusion retain 80% of preanoxic contractility. The dramatic increase in measurable iron in the ischemic, but not anoxic, heart indicates that a combination of low pH and a metabolically depleted cell is necessary for iron release and subsequent injury. This is a profoundly important observation in that lack of oxygen, per se, is less damaging than the absence of "wash-out" of the metabolic "left-overs" of the ischemic insult. It is reasonable to propose that the rapid depletion of glucose and acidosis, combined with the release of iron, occurring during no-flow ischemia is in large part responsible for functional and structural damage occurring at reperfusion following ischemia.

Another relevant study in this regard, also demonstrating the ability of hypoxic perfusion to minimize postischemic injury in skeletal muscle, was carried out by Korthuis et al.[22] In this study, the authors utilize blood rather than buffer to perfuse the canine gracilis muscle and a long ischemic insult of 4 hours followed by perfusion with either fully oxygenated or deoxygenated blood. The results indicated that postischemic perfusion with oxygenated blood leads to dramatic changes in vascular permeability, which is largely prevented if perfusion is carried out with deoxygenated blood. It is of interest to note that when the ischemic muscle was first washed out with anoxic perfusate and then with oxygenated blood, the alteration in vascular permeability was as large as noted with oxygenated perfusion alone. Thus, in this experiment there was no observable benefit of the "anoxic washout" as observed in the Langendorff preparations discussed above.

Collectively, the reports discussed above[17-22] support the contention that reperfusion following a period of stasis or no-flow, as opposed to reoxygenation of an anoxic, but perfused vascular bed, is a critical prerequisite for the genesis of reperfusion injury. In summary, it appears that the magnitude of injury caused by extended anoxic perfusion, followed by reoxygenation, is relatively modest in comparison with the severe consequences of total ischemia followed by oxygenation. It appears, based on experimental work, that "wash-out" of hydrogen ion and diffusable reducing equivalents prior to the reintroduction of oxygen, may prevent release of iron, thus limiting the severity of the resulting reperfusion injury. A second conclusion that can be drawn from these studies is that endothelial cell dysfunction occurs much earlier during an ischemic insult than do significant changes in myocardial function. Clearly, alterations in coronary flow, usually referred to as "stunning" in the context of the myocardium, are likely to represent the earliest evidence of organ injury following ischemia and reperfusion. In general, it is suggested that the microvascular compartment represents the earliest, and in many instances, the most important site of significant injury caused by ischemia and reperfusion.

Although the studies discussed above provide clear evidence that iron is released from the ischemic heart, they do not identify whether the source of this iron is predominantly endothelial or parenchymal. A question which arises in this context is: Is the endothelium particularly susceptible to the effects of ischemia, or are all cell types equally susceptible to the biochemical consequences of lack of oxygen? One might assume that cells that consume larger amounts of oxygen would be more sensitive to ischemia, but these cells may also have more efficient means to protect themselves against oxidant injury. This hypothesis can, to a first approximation, be tested in culture by exposing a variety of cells, e.g., hepatocytes, endothelial cells, myocytes and epithelial cells, to an hypoxic insult followed by "reperfusion" with exposure to normal atmosphere. The viability of the cells or their functional integrity would provide a clue to sensitivity to a hypoxic challenge. The measurement of "free" iron in the cell culture medium following such an experiment would provide evidence for a potential role of iron in the damage to the cells occurring following the ischemic insult.

Studies in the Intact Animal

In the progression from cell culture to the isolated, buffer-perfused organ, the next step is represented by "in situ" studies of specific organs, e.g., studies of the skeletal muscle, kidney or liver in an otherwise intact animal. These organs are usually blood perfused and studies of this type have high degree of physiological relevance. Several worthwhile studies of this type will be discussed in this section, but for historical reasons, and for their seminal nature, experiments by Mazur and coworkers performed at Cornell University Medical College and The New York Hospital some 40 years ago will be discussed first. These studies, published in the *Journal of Biological Chemistry* and *Journal of Clinical Investigation* between 1948 and 1958, dealt with changes in iron, xanthine oxidase and purine metabolism in animals exposed to severe hemorrhagic shock. In many ways, the studies by Mazur and coworkers foreshadowed the entire field of activated oxygen chemistry and its physiological sequelae, such as reperfusion injury, that started to take shape about 20 years following this work.

The first paper from Mazur's laboratory relevant to this discussion was published in 1955.[23] Mazur had, at the time this paper was published, already worked on the biology and physiology of ferritin for several years, having published a paper dealing with the vasoactive properties of this protein in 1948.[24] The ability of physiological stress to cause the release of iron from ferritin was first demonstrated in these pioneering efforts. It was demonstrated that when horse spleen ferritin was exposed to a variety of reducing agents, such as glutathione, cysteine and ascorbate, ferrous iron was released and could be quantitated with dipyridyl, an ferrous iron chelator. Having established that a number of reducing agents could effectively release ferrous iron from ferritin, Mazur and colleagues now wanted to determine if this type of iron release could be observed in a more physiologically relevant system. For this purpose, they incubated rat liver slices under either nitrogen or oxygen, both containing 5% carbon dioxide, at 37.5°C for 2 hours. Following this incubation they noted that ferrous iron could be measured in the supernatant and that twice as much iron was released under anaerobic conditions.

Today it is easy to criticize the design of Mazur's in vitro experiments, but his results obtained in vivo are more difficult to assail. He approached this experiment in the whole animal with the following prediction based on his in vitro experiments: " . . . liver hypoxia in vivo such as that results from prolonged drastic hemorrhagic hypotension, should result in an increase in plasma iron and a simultaneous decrease of the plasma iron binding capacity". Mazur tested his hypothesis in the Wiggers' graded hemorrhage model[25] in which dogs are bled in a controlled fashion. Initially a quantity of blood equal to 2% of the dog's weight was removed. At 15 or 30 minute intervals, smaller quantities of blood were removed until a average blood pressure of 40 mmHg was achieved. This blood pressure was then maintained by removal of additional blood or infusion of plasma for up 160 minutes, at which time all the removed blood was reinfused. During the course of the experiment, small quantities of blood were removed for analysis of iron and iron-binding capacity, with the last sample withdrawn prior to reinfusion of the blood. Of the 14 dogs exposed to the hemorrhage, 9 died within 24 hours of reinfusion of the blood and another animal shortly thereafter. The authors provided detailed results from a single animal, which demonstrated a decreased plasma iron level during the early course of the hemorrhage. This is consistent with a hypoferremia associated with general acute stress and infection, a physiological phenomenon which probably was not known at that time. After the initial decline in plasma iron levels, the levels started to increase when severe hypotension had been achieved. This animal reached a level of 2.75 mg/L of iron in its last plasma sample, compared to 0.73 mg/L prior to blood loss. The average increase in plasma iron in the ten nonsurviving dogs was 287%, while in the four surviving dogs the average increase was only 171%. The transferrin saturation at the last time point averaged 90% in the nonsurvivors and 70% in the surviving animals. The results obtained by Mazur in his canine hemorrhage studies supported his theory that reduced oxygen delivery to the tissues, in particular the liver, triggers iron release, and further suggested that the magnitude of this release may be related to outcome.

Earlier experiments from this same laboratory[26] using the same model supported this hypothesis. Baez et al demonstrated that pretreatment of the dogs with an adrenergic blocking agent, dibenamine, afforded a high degree of protection against the otherwise lethal hemorrhage regimen. The mechanism proposed by Mazur was that the adrenergic blocking agent blunts peripheral vasoconstriction, thereby allowing splanchnic perfusion even during profound shock. The authors noted no significant changes in plasma iron and iron binding capacity in these animals. Thus, the perfusion achieved by pharmacologic means was sufficient to prevent the liver and gut ischemia necessary for release of iron and therefore prevented lethal injury.

Mazur's data implicated the enzyme xanthine oxidase as the critical entity responsible for the reductive release of iron from ferritin.[27,28] He noted that intravenous administration of substrates of this enzyme, e.g., xanthine and hypoxanthine, caused a transient increase in plasma iron levels in guinea pigs, rabbits and dogs. In vivo, Mazur reasoned, this phenomenon occurs during hypoxia, when metabolic breakdown of adenine trinucleotides leads to increased levels of xanthine and hypoxanthine. The ability of Mazur and his coworkers to elucidate the exact mechanism of reduction of ferritin resulting in release of ferrous iron was hampered by the fact that they were unaware of the second reaction product of xanthine oxidase, namely the superoxide anion. Despite this handicap, Mazur clearly laid the foundation for much of the subsequent research that has led to the understanding of the etiology of vascular damage following ischemia and reperfusion.

Several years following the publication of these studies, Janoff and colleagues[29] reexamined iron status following different types of shock in rabbits. These investigators confirmed Mazur's observation that plasma levels of iron increased dramatically in hemorrhagic shock, and went on to examine if this observation also held true in other types of shock. They achieved a shock state by infusing endotoxin or catecholamine in rabbits, and by occlusion of the superior mesenteric artery in dogs. In both endotoxin and catecholamine-treated rabbits plasma iron values decreased, while in rabbits with hypovolemic shock, an increase in plasma iron was observed. These results suggest that despite the global hypotension associated with the isovolemic, hypotensive insults, there is sufficient splanchnic perfusion to prevent the delocalization of iron occurring during severe hypovolemic shock. In dogs exposed to hypovolemic hypotension of 40 mmHg, plasma iron levels were increased at all time points, but became pronounced at 6 and 9 hours of global hypotension. In the experiments involving occlusion of the mesenteric artery, plasma iron levels remained lower than normal for several hours following relief of the occlusion. Severe hemoconcentration was also evident.

Thus, Janoff and colleagues confirmed the earlier observations from Mazur's laboratory that severe hypovolemic shock leads to release of significant quantities of iron into the blood. They conclude their article by stating that " . . . It is clear, however, that the extensive pile-up of ferrous iron as globulin-bound iron in the circulation, occurring during hemorrhagic shock, is not always a feature of other forms of experimental shock".

The study by Janoff et al suggested that the observed increase in plasma iron is unique to the global shock state and to the associated ischemic state of the liver. The dogs exposed to occlusion of the superior mesenteric artery, which should render the gut ischemic, failed to yield changes in plasma iron. This may suggest that a single organ ischemic insult is insufficient to cause a systemic increase in circulating iron, or alternatively, that some degree of hypoxic perfusion of the gut was sufficient to prevent severe acidosis and the accumulation of xanthine oxidase substrates in the gut, thus failing to set the stage for massive release of iron.

Ischemic Organs Studied "In situ"

Several studies dealing with isolated organs "in vivo" have provided convincing evidence that iron may be released from an ischemic organ into the vascular space. In a study of the ischemic kidney, Paller and Hedlund[30] used rats which had had one kidney removed and the second exposed to 60 minutes of warm ischemia. Inulin clearance was used to assess renal integrity during the 24 hours of reperfusion following the ischemic insult. Iron released from the kidney into the urinary space was measured by the bleomycin technique.[7,8] A ten-fold increase in reactive iron over preischemic values in the urine during the first hour of reperfusion was observed. It was also noted that infusion of exogenous iron, administered intravenously as iron-EDTA further exacerbated the ischemic injury.

A second example demonstrating the release of iron from an organ exposed to an ischemic insult is provided by Robinson and Hedlund.[31] Rats were exposed to 120 minutes of complete intestinal ischemia by clamping the root of the mesentery. In addition to this experimental group, there were two control groups, one group having simple laparotomy and a second group exposed to incision followed by exteriorization of the intestine without clamping. Plasma iron measurements were carried out on blood samples obtained prior to surgical insult, following 80 minutes of ischemia/sham/exteriorization and following 60 and 120 minutes of reperfusion. Reperfusion was initiated after 90 minutes of ischemia. The results are illustrated in Figure 8.2. In the laparotomy group, iron levels in the plasma decreased at all points, with a very substantial decrease in the group exposed to simple abdominal incision. In the animals which had their intestine exteriorized, there was a smaller decrease in plasma iron. In the ischemic group, there was no change during ischemia, but pronounced increases following 60 and 120 minutes of reperfusion were apparent. Thus, the increased level of iron

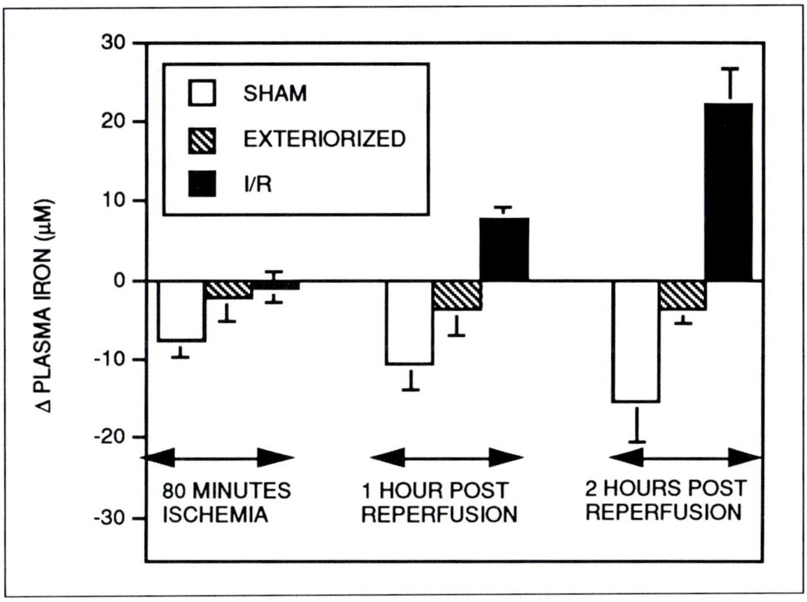

Fig. 8.2. Changes in plasma iron levels following 80 minutes of intestinal ischemia and following 60 and 120 minutes of reperfusion. The average plasma iron in the 34 animals used for this experiment was 31.2 µM prior to initiation of ischemia. Eighteen animals were used in the ischemia/reperfusion experiments and eight each in the two control groups. The increased iron level at both 60 and 120 minutes of reperfusion is statistically different from both controls at the p < 0.001 level.

observed during reperfusion more than compensates for the hypoferremia normally occurring in response to nonischemic trauma and/or infection.

In a more recent study, Chiao et al[32] examined systemic iron delocalization following partial hindlimb ischemia in the rat. An increase in transferrin saturation and a concomitant increase in plasma iron in animals exposed to ischemia and reperfusion was noted. Thus, in animals exposed only to 2 hours of hindlimb ischemia, there was a 15% increase in plasma iron compared to preischemic values. Following 2 hours of ischemia and 30 minutes of reperfusion, the plasma iron increased by 35%. Sham-treated animals (exposure of the aorta without clamping) exhibited a 15% decrease in plasma iron. These results parallel those obtained by Robinson and Hedlund[31] discussed above, suggesting that the phenomenon of iron delocalization may represent a generalized response to ischemic injury of varied etiology. Furthermore, this response appears to be common to tissues from disparate organ systems.

In summary, then, the phenomenon of iron delocalization has been demonstrated to occur in response to ischemia/reperfusion of heart and skeletal muscle, as well as of kidney and intestine. It is particularly noteworthy in models involving global ischemia, in which it appears that a significant fraction of plasma iron derives from the ischemic liver. Several studies also indicate that if iron release is prevented by metabolic washout, e.g., hypoxic/anoxic perfusion, or if iron is sequestered quickly following release, the injurious effect of reoxygention can be attenuated. Based on a significant number of studies dealing with the role of iron in reperfusion injury in disparate organ systems, it is an inescapable conclusion that free or "reactive" iron plays an important role in the genesis of reoxygenation injury and that iron sequestration, by means of targeted or systemic iron chelation represents a straightforward approach toward minimizing this type of microvascular injury.

THERAPEUTIC INTERVENTION FOR ATTENUATION OF IRON-MEDIATED INJURY

Therapeutic Intervention Targeting Iron

The concept of specific iron-targeted therapeutic measures, as a means of limiting reperfusion-mediated tissue injury, has been reviewed in a variety of publications.[33-39] The vast majority of studies that have targeted attenuation of iron mediated oxidant injury have utilized the clinically approved drug, deferoxamine (Desferal, Ciba-Geigy, Ltd., Basel, Switzerland). Of course, a much larger number of studies have used other, more general antioxidant therapeutic modalities to minimize reoxygenation injury. In many of these studies, iron is not identified as an entity that may be involved in the genesis of this injury. However, as noted above, SOD, the enzyme that converts superoxide to hydrogen peroxide and water, may well exert its protective effect by reducing superoxide mediated release of iron from its normal storage sites. This interplay between antioxidants, each with different molecular mechanisms for inhibiting oxygen mediated tissue injury, is discussed in many of the reviews listed above.

Deferoxamine

One of the earliest studies indicating that iron chelators protect an ischemic organ against ischemia and reperfusion induced injury was carried out by Hernandez and coworkers[40] using a feline intestinal model. In this model, the investigators observed protection of the intestinal microvasculature both with DFO and, more surprisingly, with apotransferrin. The end point used by these authors was intestinal capillary permeability. Comparison of the two protective agents, both given intravenously during the last 5 minutes of the 60 minute ischemic interval and for 15 of the 30 minute reperfusion interval. Deferoxamine, which freely diffuses into the extravascular compartment, infused at a dose of 50 mg/kg, may reach transient vascular concentrations of 300–400 μM, which would rapidly decrease following the end of the infusion. On the other hand, 200 mg of transferrin, restricted to the vascular bed, will be present at a concentration of 1 mg/ml of blood. Assuming a vascular volume of 200 ml in these animals, this corresponds to an iron binding capacity of 25 μM, or roughly a doubling of "normal" plasma iron binding capacity. In a separate study, Smith and colleagues[41] examined the ability of DFO and other antioxidants to prevent gastric mucosal injury following 30 minutes of ischemia (achieved by hemorrhage) followed by reperfusion (by reinfusion of shed blood). As in the case of the feline intestinal ischemia model, 50 mg/ml DFO afforded excellent protection against clearance of radioactively labeled red cells across the gastric mucosa.

In a subsequent study evaluating the role of iron in postischemic microvascular injury,[42] pretreatment with free DFO prevented alteration in postreperfusion solvent drag reflection coefficient, a measure of microvascular permeability. Pretreatment with the iron-loaded compound, ferrioxamine, was ineffective. In addition, pretreatment with apotransferrin, a chemically dissimilar iron binding agent, yielded effects similar to those of free DFO, making it exceedingly likely that DFO does in fact exert its effect through its ability to bind iron, thereby preventing catalytic activity via ferrous-ferric cycling reactions. It was also noted that the two iron binding compounds did not influence the conversion of xanthine dehydrogenase to xanthine oxidase occurring during the ischemic insult. This observation suggested that despite increased levels of xanthine oxidase during the ischemic insult, the generation of toxic radicals by this enzyme can be attenuated by the iron chelators.

The earliest study of the effect of DFO on reducing myocardial dysfunction following a 15 minute left anterior descending coronary artery (LAD) occlusion was published by Bolli et al.[43] In this pretreatment model, DFO (10 mg/kg) was given over 45 minutes starting 30 minutes prior to ischemia and continuing for the entire 15 minutes of occlusion. Another 7 mg/kg was given during the 4 hours of reperfusion. Dogs treated with the chelator showed a statistically significant improvement in the systolic thickening fraction, a direct measure of postocclusion functional recovery, as compared to the saline-treated control group at all time points during reperfusion. In a similar study, Farber et al[44] administered DFO intraatrially at a dose of 500 mg (18–26 mg/kg) to dogs subjected to 15 minutes of LAD occlusion. Infusion of the drug was initiated 15 minutes prior to LAD occlusion and was continued throughout the ischemic period. Thus, the intraatrial dose used by these authors was slightly higher than the intravenous dose given by Bolli and colleagues. Furthermore, the latter study continued slow infusion of DFO during reperfusion while the former study discontinued administration at the time of reperfusion. Despite these differences in delivery of the drug, the net recovery in thickening fraction was remarkably similar in both studies.

In a more recent study, Bolli and coworkers further evaluated the ability of DFO to improve poststunning function in the canine myocardium.[45] In this study, DFO was administered by intracoronary infusion, presumably to maximize the protective effect of the drug on the myocardial vasculature. To assess when chelator therapy was most efficacious, the authors compared the effect of infusing DFO at 3 mg/kg either immediately before or immediately after initiation of reperfusion, with low dose infusions continued throughout the reperfusion period. In addition, myocardial function in these animals was compared to animals receiving either ferrioxamine (FO, iron saturated DFO) or saline. The authors observed a remarkable recovery of function in the animals receiving DFO before, but not after, the initiation of reperfusion. The recovery of contractile function amounted to between 80 and 90%, nearly twice as large as the improvement noted in the two prior studies. There was no significant improvement in the animals receiving DFO one minute after initiation of reperfusion, nor in animals receiving FO. The authors further demonstrated, using spin trap techniques, that a burst of reactive molecules appeared in the coronary perfusate at reperfusion and that the amount of these reactive compounds was greatly reduced in animals receiving DFO prior to reperfusion. An important conclusion that can be drawn from this study is that the vascular concentration of DFO *at the beginning of reperfusion* is of critical importance in determining the efficacy of the drug.

Other Low Molecular Weight Iron Chelators

Compared to the extensive literature dealing with DFO as an antioxidant, studies utilizing other iron chelating agents are relatively sparse. However, two such studies deserve mention. Both utilize the isolated Langendorff preparation. A study by van der Kraaij and colleagues from Rotterdam demonstrated that perfusion of isolated hearts with a 50 μM solution of 1,2-dimethyl-3-hydroxy-4-pyridone (L1) afforded significant protection against a warm ischemic insult of 15 minutes.[46] Unfortunately, these authors did not directly compare the protective ability of L1 with that of DFO. A more recent study by Collis et al[47] provides an interesting comparison between the ability of a novel compound, N-methyl hexanoylhydroxamic acid, DFO and N-acetylcysteine to protect the isolated heart against reperfusion injury. The hexanyolhydroxamic acid is a hydrophobic mono-hydroxamate, while DFO can be considered a hydrophilic tri-hydroxamate. It was noted that the monohydroxamate afforded better protection than both DFO and N-acetylcysteine when used only during reperfusion. However, when used both prior to and after the ischemic insult, both hydroxamate compounds afforded considerable protection. The two novel compounds used in these studies do not appear to be superior to DFO in terms of degree of protection. Furthermore, extrapolation from results obtained in the Langendorff preparation to any form of in vivo studies is dangerous, in that systemic toxicity of novel compounds often represents a major stumbling block.

Polymer Bound Deferoxamine

Deferoxamine has received considerable attention as a powerful antioxidant by virtue of its ability to bind ferric iron through a hexadentate complex, which is unable to participate in ferrous-ferric cycling reactions. Iron bound to DFO is therefore unable to act as a catalyst in the Fenton type and lipid peroxidation reactions.[48] It is thus chemically well suited as a potential therapeutic agent for use in protection against reoxygenation injury in a variety of clinical situations. Despite being an FDA approved drug, DFO has not been widely incorporated into clinical practice. There are two reasons that may account for the very modest penetration of DFO into clinical medicine in areas outside its normal clinical use. First, due to its small size and hydrophilic nature, the drug is rapidly excreted and has a half-life in the vascular compartment of only 10–12 minutes. Second, and more importantly, free DFO is not well suited for intravenous administration due to its tendency to induce hypotension, even at relatively low doses.

Both pharmacological drawbacks of DFO can be overcome by a novel high molecular weight form of this drug, which entails the chemical attachment of the drug to biocompatible polysaccharides, such as hydroxyethyl starch and dextran.[49,50] As might be expected, the biodistribution of the drug is profoundly altered following attachment to the polymer. By virtue of its size, it remains in the vascular compartment longer. Therefore, its ability to protect the vascular compartment is enhanced, while correspondingly less protection of interstitial tissue and parenchyma is afforded. As such, the conjugate is particularly well suited to provide microvascular protection against iron mediated injury, in particular oxidant injury occurring following ischemia and reoxygenation. In addition, the attachment of the drug to the polymer abolishes its hypotensive effect. The combination of low toxicity and long vascular half-life allows for administration of polymer-drug conjugate to yield millimolar vascular levels of bound DFO. The hydroxyethyl starch-DFO (HES-DFO) conjugate has been employed in over 20 preclinical studies, demonstrating that, despite the altered biodistribution of the chelator, considerable protection can be observed against oxygen radical mediated injury in diverse preclinical models.

The hemodynamic effects of intraatrial infusion of DFO and HES-DFO have been examined by Forder et al[51] in conscious, instrumented dogs. These investigators noted hypotension and tachycardia following intraatrial infusion of 50 mg/kg of free DFO over 15 minutes. In addition, a direct effect on cardiac function, which manifested itself as a decrease in thickening fraction, was noted during the infusion period. Thus, changes in cardiac function were observed in the absence of an ischemic insult, or simply by the intraatrial infusion of DFO (3.3 mg/kg/minute). Intracoronary infusion, which may be the optimal mode of administration of an agent for achieving maximal protection, is not always available as a means of delivering drugs in acute cardiac care. Maruyama et al[52] compared the ability of DFO and hydroxyethyl starch-DFO conjugate (HES-DFO) to protect the stunned myocardium when given intravenously in a pretreatment mode. Animals were treated with HES-DFO, the iron-saturated form of the chelator (HES-FO), native DFO or saline. All treatment groups received 50 mg/kg of DFO or FO equivalents over a 45-minute period starting 30 minutes before ischemia. In this study, only animals receiving HES-DFO had a significantly improved cardiac function compared to the control groups, HES-FO and saline treated animals. The degree of protection averaged about 40% during the reperfusion. The lack of efficacy of the free drug may be due to the hypotensive effect of DFO when given intravenously. In a study using pigs, Mousa and coworkers[53] evaluated the relative efficacies of DFO and HES-DFO in a model of myocardial ischemia. The authors demonstrated that 25 mg/kg of chelator given intravenously 10 minutes into the 20-minute ischemic period resulted in a significant improvement in contractile function. Both forms of the chelator also decreased the amount of tissue edema as measured by wet/dry weight ratio, with the conjugated form showing the greatest protection. Pigs treated with DFO at a dose of 25 mg/kg exhibited significant hypotension during the initial period of reperfusion.

In a study discussed in a previous section of this chapter, Paller and Hedlund[30] compared the protective effect of free DFO with that of the chelator bound to either inulin or dextran, in a model of renal ischemia/reperfusion. Based on glomerular filtration measurements, it was shown that both free DFO and inulin-DFO afforded protection while dextran-DFO failed to do so. It was concluded that the lack of protective effect of the high molecular weight species was most likely due to the relatively poor renal filtration of this compound, thus precluding delivery of compound to the primary site of reoxygenation injury. In a study of iron release from the ischemic intestine, Robinson and Hedlund[31] tested the therapeutic efficacy of 5% HES-DFO with that of an equal volume (3.0 ml/100 g body weight) of saline containing the same amount of free DFO. These compounds were administered following 80 minutes of ischemia and ten minutes before release of the intestinal clamp. The rats treated with HES-DFO exhibited a 44% mortality, while the group treated with free DFO showed an 88% mortality within 24 hours ($p < 0.05$). In two control groups, animals treated with either HES or saline without DFO demonstrated an intermediate mortality of 66%.

A recent study by Fantini and Yoshioka[54] provides evidence that skeletal muscle electrophysiological integrity following an ischemic insult can be protected using both DFO and HES-DFO. In this in vivo rat study, 90 minutes of partial hindlimb ischemia was followed by 60 minutes of reperfusion. The protective agents, as well as saline and starch control solutions, were administered prior to clamping of the infra-renal aorta. The authors used a functional end point, namely the resting membrane potential (E_m) and the degree of depolarization of this potential following ischemia and reperfusion. In addition, muscle thiobarbituric acid reactive substances

(TBARS) were measured as an index of phospholipid peroxidation. All four groups exposed to an ischemic insult exhibited a nearly 20% reduction in E_m following 90 minutes of ischemia. Both chelator-treated groups demonstrated significant repolarization of resting E_m toward preischemic values following 60 minutes of reperfusion, while animals in both control groups failed to demonstrate significant membrane repolarization. In addition, the chelator-treated groups demonstrated a marked reduction in muscle TBARS following both ischemia and reperfusion, while significant increases were noted in both saline and starch treated groups, indicating that iron binding was associated with a reduction in lipid peroxidation. It was noteworthy that a dose of conjugated DFO of only 30 mg/kg achieved a similar therapeutic effect as 60 mg/kg of the free compound. An extracellular site of DFO action, and hence of iron delocalization and hydroxyl radical generation, is suggested by the similar protective effects achieved by the free and highmolecular weight form of the drug, the latter of which is incapable of gaining access to the intracellular compartment.

The ability of HES-DFO to attenuate microvascular injury as well or better than free DFO suggest that a significant fraction of "injurious" iron can be neutralized in the vascular, i.e. extracellular, compartment. Attachment of the chelator to the polymer makes it unable to escape this compartment except by renal and hepatic clearance, which can be further controlled by adjusting the size of the carrier molecule.

General Aspects of Therapeutic Intervention

Ischemic insult followed by reoxygenation occurs routinely during most vascular surgical procedures, as well as following trauma, when regional or global blood supply is inadequate for microvascular perfusion. During elective surgery, the amount of time that an organ is allowed to remain ischemic without leading to major organ injury is normally determined by prior experience. During such situations, any added therapy aimed at attenuation of reoxygenation injury can normally be administered prior to surgical intervention. In terms of iron chelation, it is abundantly clear that the chelator needs to be present at the time of reoxygenation, but it is less clear that additional benefit can be gained by pretreatment.

In at least two studies using DFO in canine models of myocardial ischemia, some protective effect on infarct size following coronary artery occlusion was noted when DFO was used in a pretreatment mode, but not when given at reperfusion.[55,56] These studies are complicated by the fact that insufficient quantities of the free drug can be infused acutely at reperfusion without risk of significant toxicity. This problem is avoided by the novel high molecular weight form of the chelator, which can be used routinely as part of hemodilution. In general, it seems that if there is no toxicity or other reason not to have the chelator present during the ischemic insult occurring during elective surgery, such pretreatment is likely to provide an additional benefit.

Studies on the canine myocardium also demonstrate the importance of targeting the iron chelator to the site of microvascular injury. Thus, several studies suggest that intraatrial or intracoronary infusion of the protective agent will provide a higher concentration of the chelator in the region at risk. In the case of DFO, it is difficult to achieve high vascular concentrations of the drug, since it rapidly diffuses into the extravascular space. As the novel polymer bound form of DFO has a longer vascular half-life and achieves vascular concentrations in the millimolar range following systemic administration, the need for targeted delivery may be circumvented.

One other aspect of the etiology of reoxygenation injury occurring during elective surgery deserves consideration. The studies on the isolated heart from Tsao and Lefer[19] and Voogd et al[20] discussed above clearly indicate that anoxic flow is significantly less injurious to the organ than total ischemia. These studies suggest that any form of "metabolic washout" of ischemic microvasculature at risk prior to completion of macrovascular repair, may greatly decrease subsequent reoxygenation injury. This type of metabolic washout would not require blood perfusion, i.e., is not intended to "treat" the hypoxia per se, but rather wash out hydrogen ion and metabolic products that are involved in the reductive release of iron from intracellular stores. Such washout should limit the magnitude of microvascular injury at the time of complete reperfusion and allow for improved perfusion of tissue distal to the site of macrovascular repair procedures.

Treatment of reoxygenation injury following trauma, whether this acute intervention involves limb reattachment or fluid resuscitation following severe hemorrhage, affords an excellent opportunity to "treat" the microvascular injury at the time of initial reoxygenation by incorporating the antioxidant in the first fluid reaching ischemic tissue. As noted by several investigators, particularly the studies by Bolli and colleagues[43,45] using the stunned myocardium in dogs, there is virtually no "therapeutic window" for treatment of reoxygenation injury. A large fraction of the reactive radical species are formed in the first minute following reoxygenation, suggesting that perfusion with conventional fluids (lactated Ringer's, colloid or blood) followed by antioxidant therapy is not a viable approach toward microvascular rescue. Therefore, in the context of therapeutic intervention aimed at minimizing reperfusion injury, it has been suggested that the oft-used term "the golden hour of resuscitation" coined by R. Adams Cowley, represents too generous a timeframe, and should be replaced with the more appropriate term "the golden minute of microvascular rescue".[58]

References

1. Haber F, Weiss J: The catalytic decompensation of hydrogen peroxide by iron salts. Proc. Roy. Soc. Lond. 1934;147:332-351.
2. Goldstein S, Meyerstein D, Czapski G. The Fenton reagents. Free Radic Biol Med 1993; 15:435-445.
3. Aust SD, Svingen BA: The role of iron in enzymatic lipid peroxidation, in Pryor WA, (ed): Free Radicals in Biology. New York, Academic Press, Inc., 1982, pp 1-28.
4. van Asbeck BS: Oxygen toxicity: role of hydrogen peroxide and iron. Adv Exp Med Biol 1990;264:235-246.
5. Halliwell B, Gutteridge JMC: The importance of free radicals and catalytic metal ions in human diseases. Molec Aspects Med 1985;8:89-193.
6. Minotti G. Sources and role of iron in lipid peroxidation. Chem Res Toxicol 1993; 6:134-146.
7. Gutteridge JMC, Rowley DA, Halliwell B. Superoxide-dependent formation of hydroxyl radicals in the presence of iron salts. Detection of "free" iron in biological systems by using bleomycin dependent degradation of DNA. Biochem J 1981; 199:263-265.
8. Gutteridge JMC, Rowley DA, Halliwell B. Superoxide-dependent formation of hydroxyl radicals in the presence of iron salts. Detection of "catalytic" iron and anti-oxidant activity in extracellular fluids. Biochem J 1982; 206:605-609.
9. Williams DM, Lee GR, Cartwright GE. The role of superoxide anion radical in the reduction of ferritin iron by xanthine oxidase. J Clin Invest 1974; 53:665-667
10. Biemond P, van Eijk HG, Swaak AJG et al. Iron mobilization from ferritin by superoxide derived from stimulated polymorphonuclear leukocytes. J Clin Invest 1984; 73:1576-1579.
11. Funk F, Lenders JP, Crichton RR et al. Reductive mobilization of ferritin iron. Eur J Biochem 1985; 152:167-172
12. Reif DW. Ferritin as a source of iron for oxidative damage. Free Radic Biol Med 1992; 12:417-427.
13. Halliwell B, Gutteridge JMC. Biologically relevant metal ion-dependent hydroxyl radical generation - an update. FEBS Lett 1992; 1:108-112.
14. Hebbel RP, Eaton JW. Pathology of heme interactions with the erythrocyte membrane. Sem Hematol 1989; 26:136-149.
15. Sadrzadeh SMH, Anderson DK, Panter SS et al. Hemoglobin potentiates central nervous system damage. J Clin Invest 1987; 79:662-664.
16. Gannon DE, Varani J, Phan SH et al. Source of iron in neutrophil-mediated killing of endothelial cells. Lab Invest 1987; 57:37-44.

17. Paller MS, Hedlund BE. Protection of proximal tubule epithelial cells from hypoxia/reoxygenation by extracellular iron chelators. J Am Soc Nephrol 1992; 3:711
18. Hearse DJ. Reperfusion of the ischemic myocardium. J Mol Cell Cardiol 1977; 9:605-616.
19. Tsao PS, Lefer AM. Time course and mechanism of endothelial dysfunction in isolated ischemic- and hypoxic-perfused hearts. Am J Physiol 1990; 259 (Heart Circ Physiol 28):H1660-H1666.
20. Voogd A, Sluiter W, van Eijk HG et al. Low molecular weight iron and the oxygen paradox in isolated rat hearts. J Clin Invest 1992; 90:2050-2055
21. van der Kraaij AMM, Mostert LJ, van Eijk HG et al. Iron-load increases the susceptibility of rat hearts to oxygen reperfusion damage. Protection by the antioxidant (+)-Cyanidanol-3 and deferoxamine. Circulation 1988; 78:442-449.
22. Korthuis RJ, Smith JK, Carden DL. Hypoxic reperfusion attenuates postischemic microvascular injury. Am J Physiol 1989; 256 (Heart Circ Physiol 25):H315-H319.
23. Mazur A, Baez S, Shorr E. The mechanism of iron release from ferritin as related to its biological properties. J Biol Chem 1955; 213:147-158.
24. Mazur A, Shorr E. Hepatorenal factors in circulatory homeostasis. IX. The identification of the hepatic vasodepressor substance, VDM, with ferritin. J Biol Chem 1948; 176:771-787.
25. Wiggers CJ, Werle JM. Exploration of a method for standardization of hemorrhagic shock. Proc Soc Exp Biol Med 1942; 49:604-606.
26. Baez S, Zweifach BW, Shorr E. Effect of dibenamine on humoral and vascular changes in dogs subjected to hemorrhagic shock. Fed Proc 1952; 11: 177
27. Green S, Mazur A. Relation of uric acid metabolism to release of iron from ferritin. J Biol Chem 1957; 227:653-668.
28. Mazur A, Green S, Saha A et al. Mechanism of release of ferritin iron "in vivo" by xanthine oxidase. J Clin Invest 1958; 37: 1809-1817.
29. Janoff A, Zweifach BW, Shapiro LR. Levels of plasma-bound iron in experimental shock in the rabbit and dog. Am J Physiol 1960; 198:1161-1165.
30. Paller MS, Hedlund BE. Role of iron in postischemic renal injury in the rat. Kidney Int 1988; 34:474-480.
31. Robinson E, Hedlund BE. Role of iron in ischemia and reperfusion. Circ Shock 1989; 27:367(abstract).
32. Chiao JJC, Kirschner RE, Fantini GA. Iron delocalization occurs during ischemia and persists upon reoxygenation of skeletal muscle. J Lab Clin Invest 1994 (in press).
33. Halliwell B, Gutteridge JMC. Oxygen free radicals and iron in relation to biology and medicine: Some problems and concepts. Arch. Biochem Biophys 1986; 246:501-514.
34. Halliwell B. and Gutteridge JMC. Iron toxicity and oxygen radicals. Baillières Clin Haemat 1989; 2:195-256.
35. Ward PA, Warren JS, Till GO, et al, Modification of disease by preventing free radical formation: a new concept in pharmacological intervention. Baillières Clin. Hematol 1989; 2:391-402.
36. Reilly PM, Schiller HJ, Bulkley GB. Pharmacologic approach to tissue injury mediated by free radicals and other reactive oxygen metabolites. 1991 Am J Surg 191; 161:488-500.
37. Lesnefsky, EJ. Reduction of infarct size by cell-permeable oxygen metabolite scavengers. Free Radic Biol Med 1992; 12: 429-446.
38. Rice-Evans CA, Diplock AT. Current status of antioxidant therapy. Free Radic Biol Med 1993; 15:77-96.
39. Hallaway PE, Hedlund BE. Therapeutic strategies to inhibit iron-catalyzed tissue damage. In: Lauffer R. ed. Iron and Human Disease, Boca Raton, Florida, CRC Press, 1993 pp. 477-508
40. Hernandez LA, Grisham MB, Granger DN. A role for iron in oxidant-mediated ischemic injury to intestinal microvasculature. Am J Physiol 1987; 253 (Gastrointest Liver Physiol 16): G49-G53.
41. Smith SM, Grisham MB, Manci EA, Granger DN, Kvietys PR. Gastric mucosal injury in the rat. Role of iron and xanthine oxidase. Gastroenterology 1987; 92: 950-956.

42. Smith JK, Carden DL, Grisham MB, Granger DN, Korthuis RJ. Role of iron in postischemic microvascular injury. Am J Physiol 1989; 256 (Heart Circ Physiol 25):H1472-H1477.
43. Bolli R, Patel BS, Zhu WY et al. The iron chelator DFO attenuates postischemic dysfunction. Am J Physiol 1987; 253 (Heart Circ Physiol 22):H1372-H1380.
44. Farber NE, Vercellotti GM, Jacob HS et al. Evidence for a role of iron-catalyzed oxidants in functional and metabolic stunning in the canine heart. Circ Res 1988; 63:351-360.
45. Bolli R, Patel BS, Jeroudi MO et al. Iron mediated radical reactions upon reperfusion contribute to myocardial "stunning". Am J Physiol 1990; 259 (Heart Circ Physiol 28):H1901-H1911
46. van der Kraaij AMM, Eijk HG, Koster JF. Prevention of postischemic cardiac injury by the orally active iron chelator 1,2-dimethyl-3-hydroxy-4-pyridone (L1) and the anti-oxidant (+)-cyanidanol-3. Circulation 1989;80;158-164.
47. Collis CS, Davies MJ, Rice-Evans C. Comparison of the N-methyl hexanoylhydroxamic acid, a novel antioxidant, with deferoxamine and N-acetyl cysteine against reperfusion-induced dysfunctions in the isolated rat heart. J Cardiovasc Pharmacol 1993; 22:336-342.
48. Graf E, Mahoney JR, Bryant RG et al. Iron-catalyzed hydroxyl radical formation. Stringent requirement for free iron coordination site. J Biol Chem 1984; 259:3620-3624.
49. Hallaway PE, Eaton JW, Panter SS et al. Modulation of deferoxamine toxicity and clearance by covalent attachment to biocompatible polymers. Proc Nat Acad Sci USA 1989; 86:10108-10112.
50. Mahoney JR, Hallaway PE, Hedlund BE et al. Acute iron poisoning. Rescue with macromolecular chelators. J Clin Invest 1989; 84:1362-1366.
51. Forder JR, McClanahan TB, Gallagher KP et al. Hemodynamic effects of intraatrial administration of deferoxamine or deferoxamine-pentafraction conjugate to conscious dogs. J Cardiovasc Pharm 1990; 16: 742-749.
52. Maruyama M, Pieper GM, Kalyanaraman B et al. Effects of hydroxyethyl starch conjugated DFO on myocardial functional recovery following coronary occlusion and reperfusion in dogs. J Cardiovasc Pharmacol 1991; 17:166-175.
53. Mousa SA, Ritger, RC, Smith RD. Efficacy and safety of deferoxamine conjugated to hydroxyethyl starch. J Cardiovasc Pharmacol 1992; 19:425-429.
54. Fantini GA, Yoshioka T. Deferoxamine prevents lipid peroxidation and attenuates reoxygenation injury in postischemic skeletal muscle. Am J Physiol 1993; 264 (Heart Circ Physiol 33):H1953-H1959.
55. Reddy BR, Kloner RA, Przyklenk K. Early treatment with deferoxamine limits myocardial ischemic/reperfusion injury. Free Radic Biol Med 1989; 7:45-52.
56. Lesnefsky EJ, Repine JE, Horwitz LD. Deferoxamine pretreatment reduces canine infarct size and oxidative injury. J Pharm Exp Ther 1990; 253:1103-1109
57. Bolli R, Jeroudi MO, Patel BS et al. Marked reduction of free radical generation and contractile dysfunction by antioxidant therapy begun at the time of reperfusion, Circ Res 1989; 65:607-622.
58. Hedlund BE, Hallaway PE. Role of iron and oxygen radicals in hemorrhage and shock. Klin Wochenschr 1991; 69: 1113-1117.

CHAPTER 9

Remote Effects: Lung Injury

John Mihran Davis

Gary A. Fantini

Acute lung injury as a remote sequela of severe lower torso ischemia-reperfusion (I/R) has been demonstrated experimentally.[1-5] It is the combination of ischemia and reperfusion at a site remote from the lung which results in pulmonary injury. Ischemia alone, even for prolonged periods of time, does not result in systemic injury in the absence of subsequent reperfusion.[6] Experimental models of I/R injury have identified the neutrophil as playing a central role in mediating peripheral tissue injury including skeletal muscle,[7-10] and leukosequestration in the pulmonary microvasculature is a consistent finding in animal studies of severe hind limb I/R.[1-3,11] Rendering animals leukopenic prior to the ischemic insult abrogates the pulmonary response.[3] It has become apparent that neutrophils act as common mediators of end organ injury by local release of reactive oxygen species and that circulating metabolites released from the reperfused tissue act as potent chemoattractants for neutrophil recruitment. Humoral agents involved in mediating the systemic inflammatory response to I/R injury of skeletal muscle and thought to participate in the pathogenesis of pulmonary injury include the arachidonate metabolites, thromboxane (Tx) and leukotrienes (LT), the complement system, and tumor necrosis factor. The potential roles of platelet-activating factor and nitric oxide (endothelium-derived relaxing factor) in this process have yet to be defined. However, along with the neutrophil, it is becoming apparent that the chemotactic leukotrienes occupy a central position in this cascade of events, and this chapter will focus on these compounds. Data from animal, as well as clinical studies will be reviewed, including some recent observations from an ongoing clinical investigation at The New York Hospital. In addition, the many parallels to the adult respiratory distress syndrome (ARDS) induced by exogenous lipopolysaccharide (LPS) will be explored.

Ischemia-Reperfusion Injury of Skeletal Muscle, edited by Gary A. Fantini, M.D.;

HISTORICAL PERSPECTIVE

The initial description of pulmonary failure following I/R injury of the lower extremities is believed to have been in 1969 by Stallone et al.[12] Four years later, "sequential system failure" was recognized in a series of 18 patients who initially survived surgery for ruptured aortic aneurysm (AAA).[13] All patients developed acute renal failure requiring hemodialysis, and only one survived (94% mortality rate). All patients who died developed pulmonary dysfunction necessitating ventilatory support, the pathophysiology of which was not appreciated at that time. Subsequent to this report, Baue described the clinical entity of multiple system organ failure (MSOF) in patients following traumatic injury, in a report which has been widely quoted.[14] While LPS as an etiology of MSOF has been extensively researched and well documented,[15] only recently has attention been refocused on an extrapulmonary source of I/R injury as an inciting event in the genesis of subsequent lung failure. In either event, the resulting clinical syndrome is consistently characterized by hypoxemia refractory to oxygen therapy, pulmonary edema in the presence of normal left-sided cardiac filling pressures, progressive decreases in pulmonary compliance, ventilation-perfusion mismatching and radiographic evidence of diffuse, bilateral infiltrates.[16] The association of gram-negative bacteremia with increased pulmonary capillary permeability has been recognized for over 20 years,[17] and quantifiable increases in extravascular lung water are a hallmark of the clinical syndrome.[18] The mortality rate in previously healthy individuals developing ARDS remains in excess of 65%, and survivors are frequently left with severe pulmonary sequelae.[19] As was true at the time of Stallone's report in 1969, treatment for this devastating syndrome remains largely supportive. A recent report suggests that the development of microproteinuria following AAA surgery may identify patients at increased risk for the development of pulmonary and systemic complications of remote I/R.[20]

THE COMPLEMENT SYSTEM

Historically, the role of complement has received considerable attention in the pathogenesis of ARDS. Activation of the complement pathway has been associated with the profound granulocytopenia and hypoxemia which accompanied cellophane membrane hemodialysis.[21-23] These observations initiated a theory indicting complement activation as the important initial step common to ARDS from all causes. This seemed plausible since endotoxemia, soft tissue trauma and pancreatitis all place a patient at risk for development of ARDS and all share the ability to activate complement.[24] In support of this concept, C5a, a peptide product of activated complement, was shown to have chemotactic properties and was thought to cause the intrapulmonary leukostasis seen with ARDS. Hammerschmidt et al reported in 1980 that the presence of C5a in plasma accurately identified the patients at risk for development of ARDS.[25] However, subsequent studies by Weinberg and colleagues demonstrated that complement activation is universal in patients with the sepsis syndrome.[26] More recently, Klausner and coworkers have shown that decomplementation with cobra venom factor does not inhibit the lung injury seen following I/R, using a chronic lymph fistula model in sheep.[4] However, using soluble human complement receptor type 1 (sCR1) to selectively inhibit complement, Lindsay and colleagues have provided evidence in a rat model of bilateral hind limb I/R that complement may contribute to increased lung microvascular permeability, without impacting upon pulmonary leukosequestration.[27] In any event, plasma C5a is no longer considered to be of significant value in identifying patients at risk for ARDS.[26] This finding, combined with the increasing recognition of the tissue destructive properties of the neutrophil, signalled a shift in the focus of research on the pathogenesis of ARDS to the neutrophil. As a result, much attention has been placed on the role of biologically active lipid mediators, specifically the leukotriene class of compounds.

ROLE OF THE NEUTROPHIL

The pathologically activated neutrophil is now recognized to play a critical role in various forms of tissue injury. Neutrophil mediated tissue injury has been incriminated in the pathogenesis of immune vasculitis, atherosclerosis, transplant rejection and non-steroidal anti-inflammatory drug-induced acute tubular necrosis, as well as ARDS and multiple organ failure.[28] Clearly, leuko-sequestration in the pulmonary circulation is a constant feature in experimental models of lung injury secondary to I/R of remote tissues, including intestine[29,30] and skeletal muscle.[1-3,11] Of course, neutrophils have been identified as important mediators of I/R injury in intestine[31,32] and skeletal muscle[7-10] themselves, and the process by which remote injury is effected doubtless involves an extension and amplification of shared pathophysiology.

In an attempt to causally identify the activated neutrophil in the pathogenesis of ARDS, bronchoalveolar lavage (BAL) has been employed to study the mediators and cellular elements in the alveolar air space. BAL has demonstrated the presence of neutrophil products including collagenase, myeloperoxidase elastase in patients with ARDS.[33,34] In addition, increased alveolar levels of leukotrienes have also been noted in patients with established ARDS when compared to controls.[35] Fowler and coworkers[36] prospectively evaluated three groups of patients using BAL: (1) patients at *no* risk for ARDS, (2) patients at risk for ARDS, and (3) patients with established ARDS. Lavage fluid analysis demonstrated an 8–10 fold increase in total cell count for the high risk and established ARDS groups, when compared to no risk controls. The differential counts of cells recovered by BAL revealed a 53% and 70% increase in neutrophils in the patients at risk and with ARDS respectively, and a 10-40 fold increase in protein content. The lavage fluid from patients with and at risk for ARDS also exhibited a 3-fold increase in C5a stimulated neutrophil adherence to tissue culture plastic (an indication of adherence to pulmonary artery endothelial monolayers).[37] The adherence factor was found to be composed of lipid and protein, was not neutralized by antihuman-C5a, was not present in simultaneously drawn plasma, and is now thought to be LTB_4. These studies provide indirect evidence that neutrophil influx, permeability changes and endothelial adhesion factors are present in the airways, and precede the onset of ARDS in critically ill patients.

Although a useful investigative technique, there are drawbacks with the BAL methodology that have been difficult to circumvent. For example, it has been shown in animals that mechanical ventilation and increased inspired oxygen content will alter the findings of BAL, thereby making it difficult to design studies with valid controls. In addition, ARDS is often complicated by concurrent pneumonia,[38] which would confound any subsequent observations. Finally, inflammatory cells and mediators present in the alveolar air spaces may represent a response to the predisposing factors causing ARDS, and therefore does not necessarily indicate cause and effect. Despite these shortcomings, BAL continues to be considered a valid investigative tool.

The activation of tissue macrophages are also thought to play a role in the development of ARDS. They are especially important in the severely stressed neutropenic patient. The report by Ognibene et al[39] of ARDS in neutropenic patients serves to illustrate that circulating neutrophils are not the only cells capable of producing alveolar damage. Indeed, even the profound leukopenia required for bone marrow transplant[40] does not preclude the development of ARDS. Lewis and Austen[41] have shown that polymorphonuclear leukocytes (PMNs), monocytes and macrophages produce predominantly LTB_4 in response to calcium ionophore. Resident alveolar macrophages have also been shown to produce LTB_4 as their primary neutrophil chemoattractant in response to soluble or phagocytic stimuli.[42] Furthermore, macrophages share receptor-coupled activation pathways with granulocytes, and through respiratory burst and eicosanoid production, may produce similar local injury patterns.[43] Interestingly, Rinaldo and Borovetz reported six cases of ARDS in infected, nadir, leukemia patients which became worse as their white blood cell count improved.[44] These observations serve to emphasize the important, but not exclusive, role of the activated neutrophil in the pathogenesis of acute lung injury in the clinical setting, and further serve to implicate LTB_4 as a primary mediator of this entity.

LEUKOTRIENES

The role of LT generation in critically injured patients is becoming increasingly apparent. LTB_4 and its metabolites and stereoisomers have emerged as substances likely to be responsible for the multiple and diverse physiologic disturbances seen in patients with ARDS. As early as 1938,[45] evidence of biologically active mediators existed, and in 1969 lipids were first thought to play an important role in the sepsis syndrome.[46] The identification and description of the precise pathways of leukotriene synthesis were not reported until 1976.[47,48] Since that time only a handful of publications have investigated the role of eicosanoids in clinical surgery, burns, or trauma.[49-51]

SYNTHESIS

Samuelsson's discovery of the transient intermediate LTA_4 was crucial to the elucidation of the mechanism by which cells transform arachidonic acid (AA) to leukotrienes.[47] Synthesis occurs through the release of membrane bound AA by phospholipase A-2. Addition of molecular oxygen by 5-lipoxygenase leads to the formation of 5-OH PETE, which is then converted to the unstable epoxide intermediate LTA_4. Hydration of LTA_4 leads to LTB_4. The addition of glutathione to LTA_4 by glutathione-S-transferase creates LTC_4, transpeptidases and dipeptidases then lead to LTD_4 and LTE_4.[47] Leukotriene F_4 is formed when g-glutamic acid is attached to the cysteinyl residue on LTE_4 by a transpeptidase.[4]

CYSTEINYL LEUKOTRIENES

Grouped by biological activity, LTB_4 predominantly effects leukocytes. The cysteinyl leukotrienes (LTC_4, D_4, E_4) previously known as the slow-reacting-substance of anaphylaxis (SRS-A), however, have smooth muscle as their primary target,[52] causing vasoconstriction of arterioles. They are also potent inducers of airway obstruction,[53] especially in asthmatics.[54] Moreover the cysteinyl leukotrienes increase microvascular permeability in the airways and elsewhere by their action on the endothelium of postcapillary venules.[55] When given systemically, the cysteinyl leukotrienes can induce prolonged hypotension. This phenomenon is thought to occur through capillary leak with resultant hemoconcentration, and via a direct action on the heart, reducing coronary contractility and coronary blood flow.[56-60]

EFFECT OF LTB_4 ON ENDOTHELIUM

Some evidence also exists that LTB_4 has a direct effect on vascular integrity and is produced by vascular endothelium.[61] Several investigators have demonstrated the presence of 5-lipoxygenase and the generation of peptido-leukotrienes in response to calcium ionophore in porcine pulmonary, coronary, and renal artery preparations.[62]

When LTB_4 (10^{-7}M) is added to cultured bovine pulmonary artery monolayer without neutrophils,[63] an increase in permeability is noted in association with alterations in the actin microfilaments of the cytoskeleton.[64] This increase in permeability was not significantly increased by the addition of neutrophils to the monolayer, suggesting that LTB_4 may have a direct effect on endothelial permeability. In addition, the neutrophil-mediated endothelial injury seen in vitro has been shown to be caused by the release of elastase by the activated neutrophil. This action on endothelium is greatly enhanced by the addition of small amounts of LPS, further linking gram-negative sepsis as an important etiologic factor in clinical ARDS.[65]

The local production of a potent aggregating and chemotactic substance like LTB_4 may be responsible for the initiation and amplification of the inflammatory response seen in clinical, as well as experimental, MSOF and may account for the increased sensitivity of pulmonary tissue in the pathogenesis of ARDS.

EFFECT OF LTB_4 ON THROMBOXANE GENERATION

In the rat, plasma TxB_2 levels have been shown to rise within minutes of restoration of perfusion to an ischemic limb, and a major site of Tx synthesis appears to be the lung.

Anner et al demonstrated reduced plasma TxB_2 levels, reduced pulmonary leukosequestration and normalization of lung histology in animals pretreated with a Tx synthesis inhibitor prior to severe limb ischemia.[1,2] Similarly, plasma TxB_2 levels have been reported to increase within 30 minutes of aortic clamping in patients undergoing AAA surgery, and were temporally related to intraoperative pulmonary hypertension.[66] In a subsequent report from the same group, mannitol administered prior to aortic clamping attenuated the previously observed pulmonary changes, and was found to inhibit activated platelet production of Tx in vitro.[67] However, the clinical relevance of these observations regarding Tx is uncertain, as a causal relationship has yet to be established. Indeed, subsequent animal studies by the same authors have suggested that neutrophil LTB_4 synthesis may occupy a more critical role in this regard.[68] LTB_4 has been reported to stimulate neutrophils[69] and endothelium[70] to synthesize Tx, thus the relationship between the 5-lipoxygenase and cyclooxgenase metabolites of AA is a complex one, although LTB_4 seems to occupy a much more central and important position in the cascade of events.

Effect of LTB_4 on the Neutrophil

The predominant effect of LTB_4 is on the neutrophil. Leukotriene B_4, produced largely by activated PMNs, is a powerful chemotactic and aggregating agent for mammalian neutrophils, human eosinophils, monocytes and macrophages.[71-73] It is as potent a chemoattractant as complement component C5a, and the bacterial peptide FMLP.[71,72] The adherence of leukocytes to vascular endothelium and chemotaxis to LTB_4 has been studied extensively, and is known to occur via the PMN membrane glycoprotein complex, CD18.[74-76] Both in vitro and in vivo studies have confirmed the important role of the CD18 receptor. Blockade with anti-CD18 monoclonal antibody prevents PMN accumulation and vascular leakage after subdermal injection of LTB_4,[74] and attenuates the development of multiple organ failure and death seen following hemorrhagic shock in experimental animals.[75] Further work has established that the chemotactic response is mediated by high and low affinity receptors on the PMN surface [77] occurring at concentrations as low as 10^{-10}M. In vitro, LTB_4 activates neutrophils and induces lysosomal enzyme release and toxic oxygen radical generation.[72,78,79]

At the site of injury or infection, LTB_4 promotes a number of cellular events including PMN adherence to endothelium, diapedesis, chemotaxis, and phagocytic killing. PMNs are initially directed towards opsonized bacteria by the occupation of Fc and C3b receptors on the bacterial surface. This facilitates phagocytosis and destruction of pathogenic bacteria. LTB_4 augments this process by increasing C3b receptor site expression, and enhancement of lysosomal release and superoxide generation.[79,80] Interestingly, there is a separation of the effects of LTB_4 on PMNs, along a concentration gradient. Chemotaxis and adherence[81] occur at 10- and 100-fold less LTB_4 (10^{-9}M, 10^{-8}M respectively) than the 10^{-7}M concentration required for enhancement of lysosomal release and superoxide generation. This would therefore localize destructive processes to the site of injury or infection, where the LTB_4 levels should be highest.[82]

Inadvertent extracellular release of lysosomal enzymes and reactive oxygen species at sites distant to areas of inflammation or injury may occur with severe burns, trauma, I/R and overwhelming sepsis. High circulating levels of leukotrienes and other mediators may promote generalized endothelial adhesion, PMN diapedesis, and tissue injury by activated circulating neutrophils. At this point, tissues which see a large proportion of the cardiac output and have a rich capillary network with specialized endothelium and/or resident phagocytic cells (macrophages), capable themselves of becoming activated, may be particularly susceptible to circulating and locally produced leukotrienes.

Although the understanding of the role of the circulating neutrophil in postischemic acute renal failure is still evolving,[83,84] host injury by activated PMNs has been established for ARDS.[24,40,85-90] Zimmerman et al,[86] as well as Davis and colleagues[91,92] have observed an increased state of activation of peripheral neutrophils in critically ill patients. Both prior to and after the onset of ARDS, intensive care unit patients at risk demonstrated enhanced LTB_4 production and chemotaxis, compared to control patients without ARDS,[86,91] or normal volunteers.[92] This enhanced function is similar to what is observed in neutrophils from normal volunteers infused with endotoxin,[33,93] suggesting that a priming of neutrophils occurs in response to various stimuli. It was also noted that, over time, patients with severe, persistent, unresolving pulmonary failure produced less LTB_4 than controls, and had impaired chemotaxis to LTB_4.[91] To investigate this phenomenon, normal lymphocytes and monocytes were incubated with the PMNs of patients with established ARDS.[92] Since normal PMNs coincubated with mononuclear cells exhibit synergistically enhanced LTB_4 production, it was surprising that there was no improvement in LTB_4 generation in these very ill patients. This was explained when fatty acid analysis revealed a significant (22%) reduction in the AA content of ARDS patients neutrophil membranes. The finding of reduced substrate availability was previously reported in burn patients[94] who were found to have reduced plasma levels of AA, and associated significant impairment of spontaneous random migration and chemotaxis. These findings suggest that dietary manipulation may provide a basis for future therapy.

Recently, attention has been given to the role of LTB_4 in trauma patients.[95] Acutely injured patients were prospectively studied for 8 days from the time of injury. Eicosanoid levels (LTB_4, PGE_2, PGF_2a, TxB_2 and PGI_2), hemodynamic and hematologic parameters were followed in 30 patients at risk for development of ARDS. Seven patients developed fulminant posttraumatic ARDS within 96 hours after injury. These patients, when compared to patients not developing ARDS, had a significant rise in plasma LTB_4 levels preceding the elevation of neutrophil superoxide production seen on day two. The early changes in respiratory index, pulmonary shunt, pulmonary compliance, white blood cell and platelet counts were also preceded by the rise in LTB_4, which peaked on day four. Furthermore, examination of the data from the 12 patients who became septic, using discriminant analysis, identified early LTB_4 levels together with early hematologic and hemodynamic data as significant predictors of the subsequent occurrence of sepsis. These data suggests that LTB_4 generation following traumatic injury occurs early, and may identify patients at significant risk for the development of sepsis and/or ARDS.

PULMONARY EFFECTS OF REMOTE I/R: CLINICAL INVESTIGATION

In a recent study of ten male patients who underwent successful AAA repair using graft replacement of the infrarenal aorta, analysis of neutrophil leukotriene generation, complement activation, TxB_2 generation and neutrophil oxygen consumption (an indicator of oxygen radical production) were assessed with respect to clinical outcome.[96] All operations were performed without complication related to the procedure and all patients were discharged alive from the hospital. The discharge day ranged from day 7 to day 40 postoperatively (mean 12.9 ± 3.2 days postoperatively).

Neutrophil counts more than doubled during the operative procedure. Plasma lactoferrin, a neutrophil granular enzyme assayed to document neutrophil activation, increased significantly following aortic clamping. When all data were pooled and analyzed longitudinally over the course of the surgery, no significant differences were detected in mean pulmonary arterial pressure, plasma complement levels, PMN LTB_4 generation or oxygen consumption at any time point. Additionally, mean arterial pressure and heart rate did not change significantly over the course of procedure.

While the average time from aortic clamping to the restoration of distal blood flow (removal of all vascular clamps) was 89.1 ± 8.4 minutes, two patient subgroups were identified; those with ischemic times less than 90 minutes ($n = 5$) and those with ischemic times greater than 90 minutes ($n = 5$). Pulmonary function was adversely affected by the longer aortic clamp time, as evidenced by a lower

P_aO_2/F_iO_2 (P/F) ratio in patients with longer ischemic times. In addition, these patients remained intubated longer, with 3 of 5 patients still intubated on postoperative day 2, while all patients with shorter ischemic times were extubated by the first postoperative day.

Neutrophil LTB_4 generation in eight normal controls ranged from 36.4 to 84.8 ng/10^6 PMNs, with a mean of 60.6 ng/10^6 PMNs. In patients with ischemic times of over 90 minutes, neutrophil generated LTB_4 30 minutes after removal of all vascular clamps ranged from 150 to 291 ng/10^6 PMNs and the mean, nearly four times the control value, was significantly elevated over the control level. Data from a representative patient with an ischemic time greater than 90 minutes is shown in Figure 9.1.

As noted above, the intravenous administration of mannitol, a tertiary alcohol and scavenger of hydroxyl radical, has been reported to attenuate pulmonary dysfunction in patients undergoing AAA repair.[67] Despite the uniform administration of mannitol prior to aortic clamping in the present instance, neutrophil-generated LTB_4 production and severity of pulmonary dysfunction in the postoperative period was clearly related to the duration of ischemia. However, since all patients in this series received mannitol as part of the surgical routine, it is unclear what its effects may have been. In any event, these data are consistent with the hypothesis that there is a "priming" of the neutrophil during ischemia such that LTB_4, among other mediators, is generated at reflow. The amount of LTB_4 generated appears to be linearly related to the duration of the ischemic time. Further support for this hypothesis comes from evaluation of the lactoferrin data, which indicates activation of neutrophils even as early as 5 minutes after aortic clamping. Therefore, it appears that there is a spectrum of neutrophil activation, which translates into increased LTB_4 production upon reestablishment of flow in patients with longer ischemic times. It is conceivable that exaggeration of this process may result in frank pulmonary failure, such as that observed in two patients following severe lower torso ischemia and subsequent reperfusion.[97] Precise mechanisms resulting in neutrophil activation during ischemia remain unknown. Altering neutrophil LT generation in patients undergoing aortic surgery may alter the subsequent pulmonary response, and suggests a direction for further research aimed at development of therapeutic interventions.

SUMMARY

A central role for LTB_4 in the pathogenesis of acute lung injury, secondary to either remote I/R or exogenous LPS, is suggested by several observations. Intravenous administration of LTB_4 in sheep results in a pattern of pulmonary dysfunction similar to that seen following limb I/R.[98] In addition, inhibition of LTB_4 synthesis using diethylcarbamazine prevents the pulmonary sequelae of limb ischemia.[4] LTB_4 may be generated by neutrophils, mast cells, or tissue macrophages and is known to exert potent neutrophil chemotactic effects,[71] and potentially neutrophil adhesive effects[99,100] as well. Thus, although precise mechanisms of neutrophil activation or "priming" during ischemia remain uncertain, it is clear that neutrophils are subsequently sequestered in the lung, where they exert their injurious effects culminating in pulmonary dysfunction and sometimes end organ failure. Pretreatment with antioxidants and free radical scavengers diminishes the pulmonary injury after limb ischemia experimentally, suggesting that the generation of reactive oxygen species may be the final common pathway of cell injury. Tumor necrosis factor, the generation of which may be stimulated by LTB_4 in the lung,[101] may play an important, though as yet incompletely defined role.[102] As platelet-activating factor may be released by ischemic skeletal muscle,[103] its capacity to modulate leukocyte adhesion to endothelium during I/R[104,105] suggests a role for this molecule in mediating remote injury as well. Finally, the extent to which the ubiquitous autocoid nitric oxide,[106,107] with the capacity to modulate microvascular permeability,[108] inhibit neutrophil adherence to endothelium[109] and participate in the generation of hydroxyl radical,[110] participates in the genesis of remote injury during I/R remains the subject of active investigation.

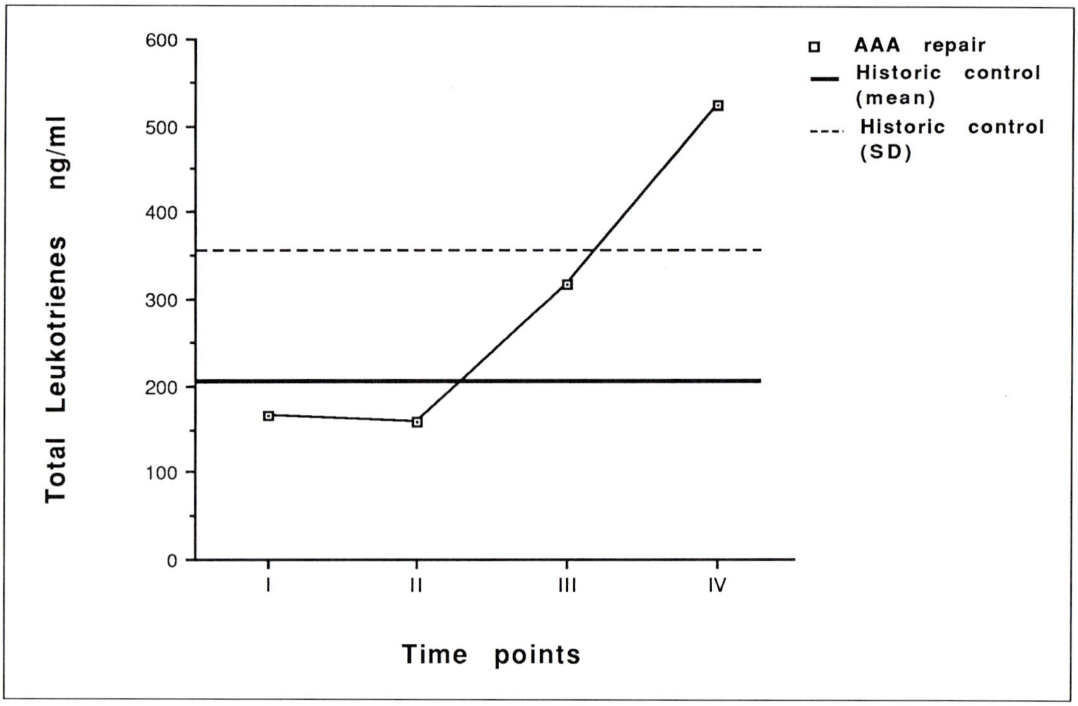

Fig. 9.1a. Depicting total chemotactic leukotrienes from a representative patient with an ischemic time of greater than 90 minutes during AAA surgery. Control values (mean ± standard deviation) were obtained from normal volunteers (n = 8) in each instance. Time points: I, after induction of general anesthesia; II, 5 minutes after application of vascular clamps; III, 5 minutes after all vascular clamps were removed; IV, 30 minutes after all vascular clamps were removed.

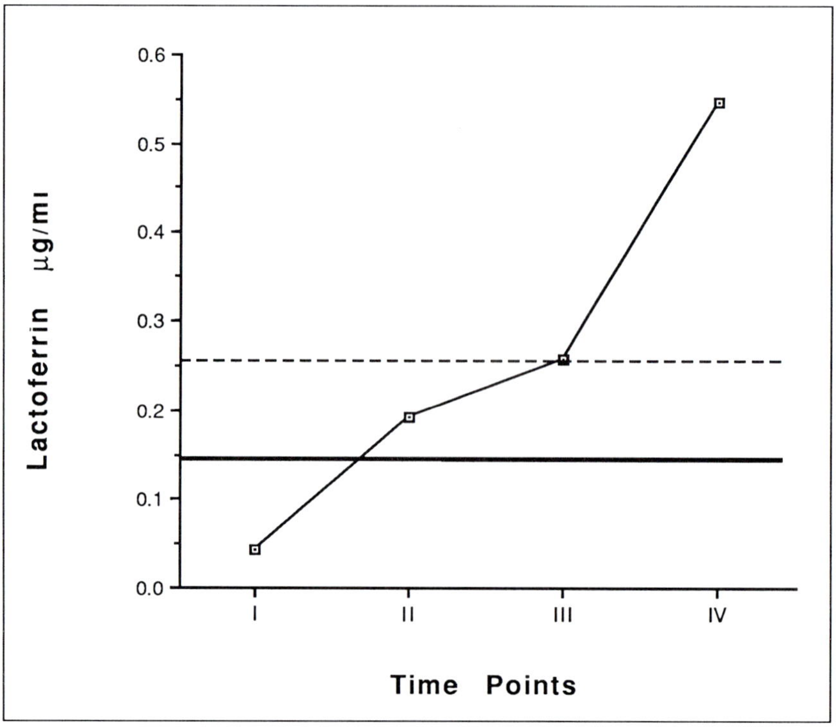

Fig. 9.1b. Representing plasma lactoferrin values.

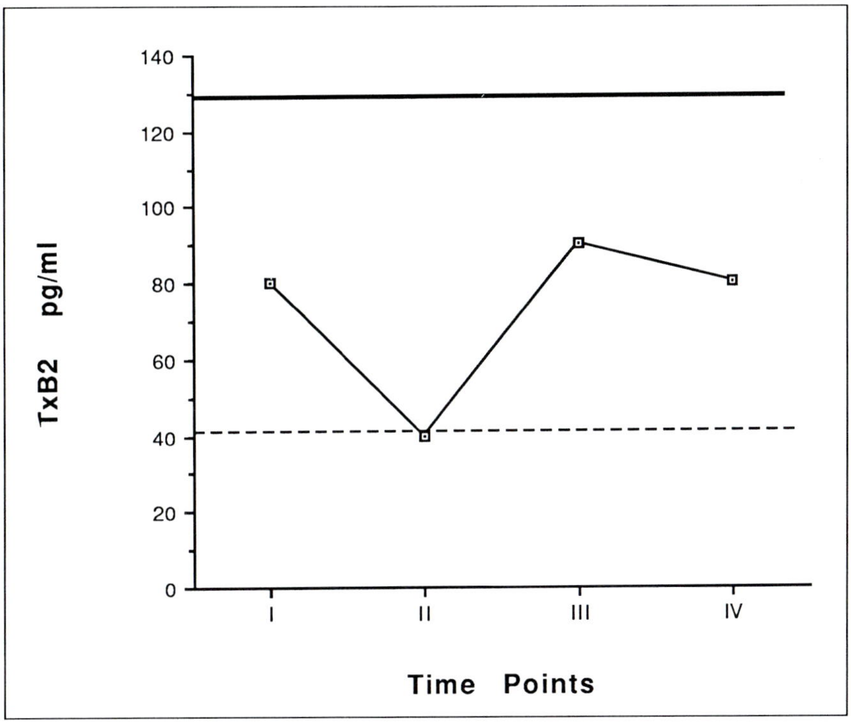

Fig. 9.1c. Representing plasma thromboxane B_2 values.

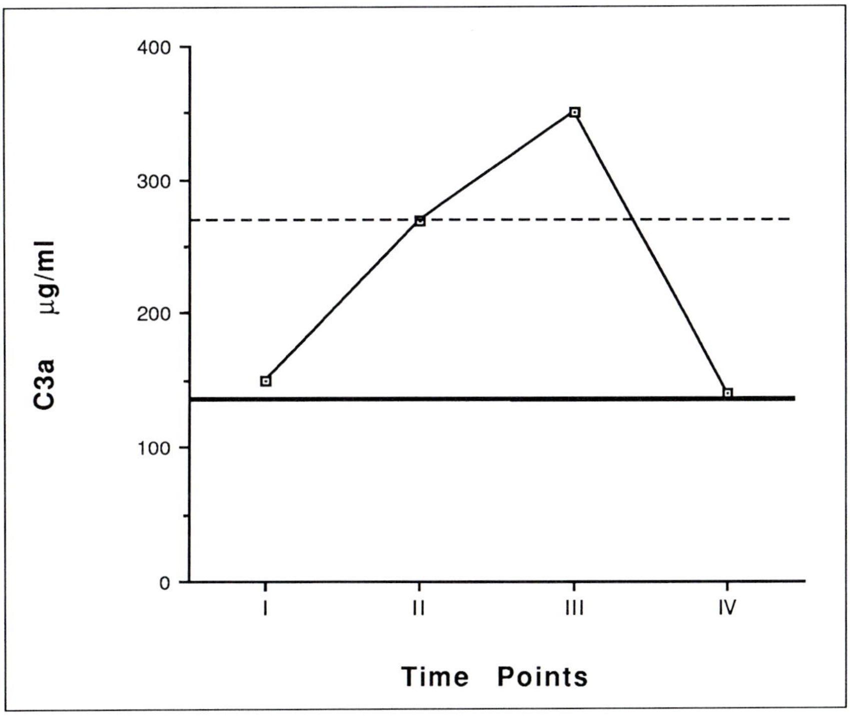

Fig. 9.1d. Representing plasma C3a levels.

References

1. Anner H, Kaufman RP, Kobzik L et al. Pulmonary leukosequestration induced by hind limb ischemia. Ann Surg 1987; 206:162-167.
2. Anner H, Kaufman RP, Kobzik L et al. Pulmonary hypertension and leukosequestration after lower torso ischemia. Ann Surg 1987; 206:642-648.
3. Klausner JM, Anner H, Paterson IS et al. Lower torso ischemia-induced lung injury is leukocyte dependent. Ann Surg 1988; 208:761-767.
4. Klausner JM, Paterson IS, Kobzik L et al. Leukotrienes but not complement mediate limb ischemia-induced lung injury. Ann Surg 1989; 209:462-470.
5. Punch J, Rees R, Cashmer B et al. Acute lung injury following reperfusion after ischemia in the hind limbs of rats. J Trauma 1991; 31:760-767.
6. Seekamp A, Ward PA: Ischemia-reperfusion injury, in Oppenheim JJ, (eds): Inflammatory Disease Therapy. Basel, Birkhauser Verlag, 1993, pp 137-152.
7. Yokota J, Minei JP, Fantini GA et al. Role of leukocytes in reperfusion injury of skeletal muscle after partial ischemia. Am J Physiol 1989; 257:H1068-1075.
8. Korthuis RJ, Grisham MB, Granger DN. Leukocyte depletion attenuates vascular injury in postischemic skeletal muscle. Am J Physiol 1988; 254(23):H823-827.
9. Rubin BB, Smith A, Liauw S et al. Complement activation and white cell sequestration in postischemic skeletal muscle. Am J Physiol 1990; 259:H525-531.
10. Belkin M, LaMorte ML, Wright JG et al. The role of leukocytes in the pathophysiology of skeletal muscle ischemic injury. J Vasc Surg 1989; 10:14-17.
11. Klausner JM, Paterson IS, Valeri CR et al. Limb ischemia-induced increase in permeability is mediated by leukocytes and leukotrienes. Ann Surg 1988; 208:755-760.
12. Stallone RJ, Lim RC, Blaisdell FW. Pathogenesis of the pulmonary changes following ischemia of the lower extremities. Ann Thorac Surg 1969; 7:539-549.
13. Tilney NL, Bailey GL, Morgan AP. Sequential system failure after rupture of abdominal aortic aneurysms: An unsolved problem in postoperative care. Ann Surg 1973; 178: 117-122.
14. Baue AE. Multiple, progressive or sequential systems failure: a syndrome of the 1970s. Arch Surg 1975; 110:779-781.
15. Montgomery AB, Stager MA, Carrico CJ et al. Causes of mortality associated with the adult respiratory distress syndrome. Am Rev Respir Dis 1985; 132:485-489.
16. Barie PS: Acute respiratory failure, in Barie PS, Shires GT, (eds): Surgical Intensive Care. Boston, Little, Brown and Company, 1993: 227-283.
17. Brigham KL, Woolverton WC, Blake LH et al. Increased sheep lung vascular permeability caused by pseudomonas bacteremia. J Clin Invest 1974; 54:792-804.
18. Tranbaugh RF, Lewis FR, Christensen JM et al. Lung water changes after thermal injury. The effects of crystalloid resuscitation and sepsis. Ann Surg 1980; 192: 479-490.
19. Fowler AA, Hamman RF, Zerbe GO et al. Adult respiratory distress syndrome. Prognosis after onset. Am Rev Respir Dis 1985; 132:472-478.
20. Smith FCT, Gosling P, Sanghera K et al. Microproteinuria predicts the severity of systemic effects of reperfusion injury following infrarenal aortic aneurysm surgery. Ann Vasc Surg 1994; 8:1-5.
21. Kaplow LS, Goffinet JA. Profound neutropenia during the early phase of hemodialysis. JAMA 1968; 203:1135-1137.
22. Craddock PR, Fehr J, Dalmasso AP et al. Hemodialysis leukopenia. Pulmonary vascular leukostasis resulting from complement activation by dialyzer cellophane membranes. J Clin Invest 1977; 59:879-888.
23. Craddock PR, Fehr J, Brigham KL et al. Complement and leukocyte-mediated pulmonary dysfunction in hemodyalisis. N Engl J Med 1977; 296:769-774.
24. Rinaldo JE. Mediation of ARDS by leukocytes. Clinical evidence and indications for therapy. Chest 1986; 89:590-593.
25. Hammerschmidt DE, Weaver LJ, Hudson LD et al. Association of complement activation and elevated plasma C5a with adult respiratory distress syndrome. Pathophysiologic

relevence and possible prognostic value. Lancet 1980; 1:947-949.
26. Weinberg PF, Matthay MA, Webster RO et al. Biologically active products of complement and acute lung injury in patients with the sepsis syndrome. Am Rev Respir Dis 1984; 130:791-796.
27. Lindsay TF, Hill J, Ortiz F et al. Blockade of complement activation prevents local and pulmonary albumin leak after lower torso ischemia-reperfusion. Ann Surg 1992; 216:677-683.
28. Anderson BO, Brown JM, Harken AH. Mechanisms of neutrophil-mediated tissue injury. J Surg Res 1991; 51:170-179.
29. Hill J, Lindsay T, Valeri CR et al. A CD18 antibody prevents lung injury but not hypotension following intestinal ischemia-reperfusion. J Appl Physiol 1993; 74: 659-664.
30. Schmeling DJ, Caty MG, Oldham KT et al. Evidence for neutrophil-related acute lung injury after intestinal ischemia-reperfusion. Surgery 1989; 106:195-202.
31. Grisham MB, Benoit JM, Granger DN. Assessment of leukocyte involvement during ischemia and reperfusion of intestine. Methods Enzymol 1990; 186:729-743.
32. Zimmerman BJ, Guillory DJ, Grisham MB et al. Role of leukotriene B_4 in granulocyte infiltration into postischemic feline intestine. Gastroenterology 1990; 99:1358-1363.
33. Lee CT, Fein AM, Lippmann M et al. Elastolytic activity in pulmonary lavage fluid from patients with adult respiratory distress syndrome. N Engl J Med 1981; 304: 192-196.
34. Wewers MD, Herzyk DJ, Gadek JE. Alveolar fluid neutrophil elastase activity in the adult respiratory distress syndrome is complexed to alpha-2 macroglobulin. J Clin Invest 1988; 82:1260-1267.
35. Stephenson AH, Lonigro AJ, Hyers TM et al. Increased concentration of leukotrienes in bronchoalveolar lavage fluid of patients with ARDS or at risk for ARDS. Am Rev Respir Dis 1988; 138:714-719.
36. Fowler AA, Hyers TM, Fisher BJ et al. The adult respiratory distress syndrome. Cell populations and soluble mediators in the air spaces of patients at high risk. Am Rev Respir Dis 1987; 136:1225-1231.
37. Webster RO, Wysolmerski RB, Lagunoff D. Enhancement of human polmor - phonuclear neutrophil adherence to plastic and endothelium by phorbol myristate acetate. Comparison with C5a. Am J Pathol 1986; 125:369-378.
38. Bell RC, Coalson JJ, Smith JD et al. Multiple organ system failure and infection in adult respiratory distress syndrome. Ann Intern Med 1983; 99:293-298.
39. Ognibene FP, Martin SE, Parker MM et al. Adult respiratory distress syndrome in patients with severe neutropenia. N Eng J Med 1986; 315:547-551.
40. Larsen GL, Parrish DA, Henson PM. Lung-defense: The paradox of inflammation. Chest 1983; 83 (suppl 5):15-55.
41. Lewis RA, Austen KF. The biologically active leukotrienes. Biosynthesis, metabolism, receptors, functions, and pharmacology. J Clin Invest 1984; 73:889-897.
42. Martin TR, Raugi G, Merritt TL et al. Relative contribution of leukotriene B4 to the neutrophil chemotactic activity produced by the resident human alveolar macrophage. J Clin Invest 1987; 80:1114-1124.
43. Rinaldo JE, Rogers RM. Adult respiratory distress syndrome [editorial]. N Engl J Med 1986; 315:578-580.
44. Rinaldo JE, Borovetz H. Deterioration of oxygenation and abnormal lung microvascular permeability during resolution of leukopenia in patients with diffuse lung injury. Am Rev Respir Dis 1985; 131: 579-583.
45. Feldberg W, Kelly CH. Liberation of histamine and formation of substances by cobra venom. J Physiol 1938; 94:187-226.
46. Gallin JI, Kaye D, O'Leary WM. Serum lipids in infection. N Engl J Med 1969; 281:1081-1086.
47. Samuelsson B. Leukotrienes: Mediators of immediate hypersensitivity reactions and inflammation. Science 1983; 220:568-575.
48. Borgeat P, Hamberg M, Samuelsson B. Transformation of arachidonic acid and homo-gamma-linolenic acid by rabbit polymorphonuclear leukocytes. Monohydroxy acids from novel lipoxygenases. J Biol Chem 1976; 251:7816-7820.

49. Craft DV, Lefer DJ, Hock CE et al. Significance of production of peptide leukotrienes in murine traumatic shock. Am J Physiol 1986; 251:H80-85.
50. Alexander JW, Saito H, Trocki O et al. The importance of lipid type in the diet after burn injury. Ann Surg 1986; 204:1-8.
51. Stjernschantz J. The leukotrienes. Med Biol 1984; 62:215-230.
52. Samuelsson B, Dahlen SE, Lindgren JA et al. Leukotrienes and lipoxins: Structures, biosynthesis and biological effects. Science 1987; 237:1171-1176.
53. Weiss JW, Drazen JM, Coles N et al. Bronchoconstrictor effects of leukotriene C in humans. Science 1982; 216:196-198.
54. Dahlen SE, Hansson G, Hedqvist P et al. Allergen challenge of lung tissue from asthmatics elicits bronchial contraction that correlates with release of leukotrienes C4 D4 and E4. Proc Natl Acad Sci USA 1983; 80:1712-1716.
55. Dahlen SE, Bjork J, Hedqvist P et al. Leukotrienes promote plasma leakage and leukocyte adhesion in postcapillary venules: In vivo effects with relevance to the acute inflammatory response. Proc Nat Acad Sci USA 1981; 78:3887-3891.
56. Letts LG, Piper PJ. The actions of leukotrienes C4 and D4 on guinea pig isolated hearts. Br J Pharmacol 1982; 76:169-176.
57. Kito G, Okuda H, Ohkawa S et al. Contractile activities of leukotriene C4 and D4 on vascular strips from rabbits. Life Sci 1981; 29:1325-1332.
58. Burke JA, Levi R, Guo Z et al. Leukotrienes C4 D4 and E4: Effects on human and guinea-pig cardiac preparation in vitro. J Pharmacol Exp Ther 1982; 221:235-241.
59. Michelassi F, Lando L, Hill RD et al. Leuktriene D4: A potent coronary artery vasoconstrictor associated with impaired ventricular contraction. Science 1982; 217:841-843.
60. Smedegard G, Hedqvist P, Dahlen SE et al. Leukotriene C4 affects pulmonary and cardiovascular dynamics in monkeys. Nature 1982; 295:327-329.
61. Piper PJ, Galton SA. Generation of leukotriene B4 and leukotriene E4 from porcine pulmonary artery. Prostaglandins 1984; 28:905-914.
62. Piper PJ, Letts LG, Galton SA. Generation of a leukotriene-like substance from porcine pulmonary artery. Prostaglandins 1983; 25:591-599.
63. Malik AB, Noonan TC, Selig WM et al. Effects of exogenous leukotrienes on the pulmonary circulation, in Lefer AM, Gee MH, (eds): Leukotrienes in Cardiovascular and Pulmonary Function. New York, A R Liss, 1985, pp 221-235.
64. Wells JS, Shepro D, Hechtman HB. Eicosanoid modulation of stress fibers in cultured bovine aortic endothelial cells. Inflammation 1985; 9:439-450.
65. Smedly LA, Tonnesen MG, Sandhaus RA et al. Neutrophil-mediated injury to endothelial cells. Enhancement by endotoxin and essential role of neutrophil elastase. J Clin Invest 1986; 77:1233-1243.
66. Paterson IS, Klausner JM, Pugatch R et al. Noncardiogenic pulmonary edema after abdominal aortic aneurysm surgery. Ann Surg 1989; 209:231-236.
67. Paterson IS, Klausner JM, Goldman G et al. Pulmonary edema after aneurysm surgery is modified by mannitol. Ann Surg 1989; 210:796-801.
68. Goldman G, Welbourn R, Paterson IS et al. Ischemia-induced neutrophil activation and diapedesis is lipoxygenase dependent. Surgery 1990; 107:428-433.
69. Puustinen T, Uotila P. Thromboxane formation in human polymorphonuclear leukocytes is inhibited by prednisone and stimulated by leukotrienes B4, C4 and histamine. Prostaglandins Leukot Med 1984; 14:161-167.
70. Dunham S, Shepro B, Hechtman HB. Leukotriene B4 induction of TXB2 in cultured bovine aortic endothelial cells. Inflammation 1984; 8:312-321.
71. Ford-Hutchinson AW, Bray MA, Doig VM et al. Leukotriene B, a potent chemokinetic and aggregating substance released from polymorphonuclear leukocytes. Nature 1980; 286:264-265.
72. Bray MA. The pharmacology and pathophysiology of leukotriene B4. Br Med Bull 1983; 39:249-254.
73. Thureson-Klein A, Hedquist P, Lindbom D. Ultrastructure of polymorphonuclear leukocytes in postcapillary venules after exposure

to leukotriene B4 in vivo. Acta Physiol Scand 1984; 122:221-224.

74. Arfors KE, Lundberg C, Lindbom L et al. A monoclonal antibody to the membrane glycoprotein complex CD 18 inhibits polymorphonuclear leukocyte accumulation and plasma leakage in vivo. Blood 1987; 69:338-340.
75. Vedder NB, Winn RK, Rice CL et al. A monoclonal antibody to the adherence-promoting leukocyte glycoprotein, CD 18, reduces organ injury and improves survival from hemorrhagic shock and resuscitation in rabbits. J Clin Invest 1988; 81:939-94.
76. Welbourn R, Goldman G, Kobzik L et al. Neutrophil adherence receptors (CD 18) in ischemia. Dissociation between quantitative cell surface expression and diapedesis mediated by leukotriene B4. J Immunol 1990; 145:1906-1911.
77. Goldman DW, Goetzl EJ. Specific binding of LTB4 to receptors on human polymorphonuclear leukocytes. J Immunol 1982; 129:1600-1605.
78. Shoell HJ, Naccache PH, Borgeat P et al. Characterization of the secretory activity of LTB4 toward rabbit PMN's. J Immunol 1982; 128:811-816.
79. Serhan CN, Smolen JE, Korchak HM et al. Leukotriene B4 is a complete secretagogue in human neutrophils: Ca^{2+} translocation in liposomes and kinetics of neutrophil activation. Adv Prostaglandin Leukot Res 1983; 11:53-63.
80. Nagy L, Lee TH, Goetzl EJ et al. Complement receptor enhancement and chemotaxis of human neutrophils and eosinophils by leukotrienes and other lipoxygenase products. Clin Exp Immunol 1982; 47:541-547.
81. Gimbrone MA, Brock AF, Schaefer AL. Leukotriene B4 stimulates polymorphonuclear leukocyte adhesion to cultured vascular endothelial cells. J Clin Invest 1984; 74:1552-1556.
82. Sprague RS, Stephenson AH, Dahms TE et al. Proposed role for leukotriense in the pathophysiology of multiple system organ failure. Crit Care Clin 1989; 5:315-329.
83. Klausner JM, Paterson IS, Goldman G et al. Postischemic renal injury is mediated by neutrophils and leukotrienes. Am J Physiol 1989; 256:F794-802.
84. Thornton MA, Winn R, Alpers CE et al. An evaluation of the neutrophil as a mediator of in vivo renal ischemic-reperfusion injury. Am J Pathol 1989; 135:509-515.
85. Tate RM, Repine JE. Neutrophils and the adult respiratory distress syndrome. Am Rev Respir Dis 1983; 128:552-559.
86. Zimmerman GA, Rennzetti AD, Hirc HR. Functional and metabolic activity of granulocytes from patients with adult respiratory distress syndrome: Evidence for activated neutrophils in the pulmonary circulation. Am Rev Respir Dis 1983; 127:290-300.
87. Dowe JE, Short A, Sibbald WJ et al. Pulmonary accumulation of polymorphonuclear leukocytes in the adult respiratory distress syndrome. Crit Care Med 1982; 10:712-718.
88. Shasby DM, Vanbenthuysen KM, Tate RM et al. Granulocytes mediate acute edematous lung injury in rabbits and in rabbit lungs perfused with phorbol myristate acetate: Role of oxygen radicals. Am Rev Respir Dis 1982; 125:443-447.
89. Repine JE, Cheronis JC, Rodeell TC et al. Pulmonary oxygen toxicity and ischemia-reperfusion injury: A mechanism in common involving xanthine oxidase and neutrophils. Am Rev Respir Dis 1987; 136:483-485.
90. Welbourn CRB, Goldman G, Paterson IS et al. Pathophysiology of ischemia reperfusion injury: Central role of the neutrophil. Br J Surg 1991; 78:651-655.
91. Davis JM, Meyer JD, Barie PS et al. Elevated production of neutrophil leukotriene B4 precedes pulmonary failure in critically ill surgical patients. Surg Gynecol Obstet 1990; 170:495-500.
92. Davis JM, Yurt RW, Barie PS et al. Leukotriene B4 generation in patients with established pulmonary failure. Arch Surg 1989; 124:1451-1455.
93. Meyer JD, Yurt RW, Duhaney R et al. Differential neutrophil activation before and after endotoxin infusion in enterally versus parenterally fed volunteers. Surg Gynecol Obstet 1988; 167:501-509.
94. Davis JM, McNamara D, Amand R et al. The relationship between chemotaxis and lipid levels in burns and normal subjects. Eur J Surg Res 1984; 16:59-60 [abstract].

95. Rivkind AI, Siegel JH, Guadalupi P et al. Sequential patterns of eicosanoid, platelet, and neutrophil interactions in the evolution of the fulminant post-traumatic adult respiratory distress syndrome. Ann Surg 1989; 210:355-373.
96. Gadaleta D, Fantini GA, Silane MF et al. Neutrophil leukotriene generation and pulmonary dysfunction following abdominal aortic aneurysm repair. Surgery In Press.
97. Conte MS, Fantini GA. Pulmonary failure following lower torso ischemia: clinical evidence for a remote effect of reperfusion injury. [submitted for publication]
98. Garcia JGN, Coonan TC, Jubiz W. Leukotrienes and the pulmonary microcirculation. Am Rev Respir Dis 1987; 136:161-169.
99. Wallis WJ, Hickstein DD, Schwartz BR et al. Monoclonal antibody-defined functional epitopes on the adhesion-promoting glycoprotein complex (CDw18) of human neutrophils. Blood 1986; 67:1007-1013.
100. Tonnesen MG, Anderson DC, Springer TA et al. Adherence of neutrophils to cultured human microvascular endothelial cells. J Clin Invest 1989; 83:637-646.
101. Goldman G, Welbourn R, Kobzik L et al. Lavage with leukotriene B4 induces lung generation of tumor necrosis factor-alpha that in turn mediates neutrophil diapedesis. Surgery 1993; 113:297-303.
102. Welbourn R, Goldman G, O'Riordain M et al. Role for tumor necrosis factor as mediator of lung injury following lower torso ischemia. J Appl Physiol 1991; 70:2645-2649.
103. Duran WN, Milazzo V, Hobson II RW. Microcirculatory dysfunction in ischemia/reperfusion. Role of platelet activating factor. J Vasc Surg 1993; 18:113-114.
104. Dillon PK, Fitzpatrick MF, Ritter AB et al. Effect of platelet activating factor on leukocyte adhesion to microvascular endothelium. Time course and dose-response relationships. Inflammation 1988; 12: 563-573.
105. Kubes P, Ibbotson G, Russel J et al. Role of platelet-activating factor in ischemia/reperfusion-induced leukocyte adherence. Am J Physiol 1990; 259:G300-305.
106. Moncada S, Higgs A. The L-arginine-nitric oxide pathway. N Engl J Med 1993; 329:2002-2012.
107. Palmer RMJ. The discovery of nitric oxide in the vessel wall. A unifying concept in the pathogenesis of sepsis. Arch Surg 1993; 128:396-401.
108. Kubes P, Granger DN. Nitric oxide modulates microvascular permeability. Am J Physiol 1992; 262:H611-615.
109. Gaboury J, Woodman RC, Granger DN et al. Nitric oxide prevents leukocyte adherence: role of superoxide. Am J Physiol 1993; 265:H862-867.
110. Beckman JS, Beckman TW, Chen J et al. Apparent hydroxyl radical production by peroxynitrite: implications for endothelial injury from nitric oxide and superoxide. Proc Natl Acad Sci USA 1990; 87:1620-1624.

INDEX

Page numbers in italics denote figures (f) and tables (t).

A

Acetylcholine, 129
Acetyltransferase, 63, 67
Adenosine, 42
Ambruso DR, 45
Anaphylatoxins, 80, 82
Antineutrophil serum, *15f*
a-1-antiprotease, 20, 48
a_1-antiproteinase inhibitor (API), 48, *50f*
Antiproteinases, 48
Apotransferrin, 12, 135
Arachidonic acid (AA), 64, 146, 148
ARDS (acute respiratory distress syndrome), 144
 ischemia-reperfusion (I/R)
  clinical investigations, 148-149
  complement, 144
  leukotrienes, 146-148
  neurophils, 144-146
L-Arginine, 103
Atherosclerosis
 diabetes mellites (DM), 110
 nitric oxide (NO), 109
ATP (adenosine triphosphate), 7
Austen KF, 145
Autoimmune disease, 114
Autoregulation, 2

B

Baez S, 132
Bagge U, 51
Barbaro JF, 61, 62, 66
Barzaghi G, 62
Baue AE, 3, 144
Bayliss W, 2
Beckman JS, 7
Benveniste J, 62
Bevilacqua MP, 94
Biemond P, 127
Bolli R, 135, 138
Borovetz H, 145
Bredt DS, 104
Brett J, 97
Bronchoalveolar lavage (BAL), 145
Bulkley GB, 3
Burn patients, 148
Burton AC, 2

C

C1 inhibitor, 80
C1q, 78, 81
C3, 87
 convertase, 79
C3a, *151f*
C3b, 147
C4, 83
 analysis, 85-87
 binding protein, 81
C5, 87
C5a, 144, 145, 147
Ca^{2+}, 20
Calcium, 2
 nitric oxide synthase (NOS), 106-107, 108
Calcium paradox, 3
Calmodulin
 nitric oxide synthase (NOS), 106-107, 108
Cambria RA, 45
Carden DL, 40
Catalase (CAT), 10, 12, 110
CD11/CD18. See also Integrin.
 no-reflow phenomenon, 21
CD11, *15f, 16f,* 16
CD18, *15f, 16f,* 16-17, 40
Chiao JJC, 134
Chronic granulomatous disease (CGD), 49
Coagulation, 98
Cobra venom factor, 81
Collagenase, 49
Compartment syndrome, 10
Complement, *14f,* 19, 46. 77, *79f*
 activation
  after tissue ischemia, 81-82
  skeletal muscle, 82-83
 alternate pathway, 79-80
  activation after experimental muscle ischemia, 83
   C4 analysis, 85-87
   factor B analysis, 85
   neutrophil sequestration, 84-85
 ARDS (acute respiratory distress syndrome), 144
 blockade, 87, 89-90
 classical pathway, 78-79, *89f*
 functions, 78
 membrane attack complex, 79
 regulation of cascade, 80

Complement receptor 1 (CR1), 81
gene, 82
soluble (sCR1), 82, 87, *90f*
ARDS (acute respiratory distress syndrome), 144
Complement receptor 3 (CR3). See Mac-1.
Complement receptor 4 (CR4). See gp150,95.
Cowley RA, 139
Cronstein BN, 42
Cyclic AMP (cAMP), 65
Cytokines, *95t, 96t*

D

Davis JM, 148
Decay accelerating factor, 81
Deferoxamine (DFO), 12, 45, 128, 134-136
polymer bound, 136-138
Deguchi T, 104
Del Maestro RF, 3
Demopoulos CA, 62
Desferal, 134
Dexamethasone, 68, 108
Diabetes mellites (DM), 110, 114
Diacylglycerol (DAG), 64
Dimethyl sulfoxide (DMSO), 12
Dimethylthiourea (DMTU), 12

E

Ecklof B, 1
Edema, 22, *22f*
Elastase, 20, 48, 49
Electron paramagnetic resonance (EPR), 109
Endothelial cells, 3
leukocyte adhesion, 13, 15, 21, 22, *22f*
leukotriene B_4 (LTB_4), 46
platelet-activating factor (PAF), 67
Endothelial-leukocyte adhesion molecule-1 (ELAM-1), 98. See also E-selectin.
Endothelium-derived relaxing factor (EDRF), 7, 104, 108-110. See also Nitric oxide.
Eosinophil, 66
Erlansson M, 35
Ernst E, 2

F

Factor B, 83, 85, *86f*
Factor H, 81
Factor I, 81
Fantini GA, 12, 137
Farber NE, 135
Fenton reaction, 7, 44, 126
Ferrioxamine (FO), 135
Ferritin, 125. 126, 127
FITC-dx150, 68
Forder JR, 137
Fowler AA, 145
Free radical scavengers, 10, *11f,* 12
Funk F, 127
Furchgott RF, 104, 109

G

Gannon DE, 45, 128
Garthewaite J, 104
Gaston B, 106
Gelatinase, 20, 49
Genistein, 65
Glutathione, 146
Glutathione peroxidase, 43
Glyceraldehyde-3-phosphate dehydrogenase, 106
cGMP (cyclic guanine monophosphate), 104, 105, 108
Goldberg M, 35
gp150,95, 38
Granger DN, 3
Granulocyte-monocyte colony stimulating factor (GM-CSF), 67
Guanylate cyclase, 105, 108

H

Haber-Weiss reaction, 7, 44, 45, 126
Harman JW, 2, 49
Harris AG, 51
Hearse DJ, 129
Heart, 2
leukocyte sequestration, 33-34
Hechtman H, 3
Hedlund BE, 128. 133, 134, 137
Heideman M, 82
Hemeprotein, 7, 127-128
Hemoglobin, 125
Hemosiderin, 125
Hepatocytes, 108, 112
Hernandez LA, 134
Hibbs Jr JB, 103
Histamine, 81
Honda ZI, 62, 64
Human umbilical vein endothelial cells (HUVEC), 40
Hwang SB, 62, 64
Hydrogen peroxide, 6, 47, 48, 114
Hydroxyethyl starch desferoxamine (HES-DFO), 137
Hydroxyl radical, 6-7, 8, 114, 126
Hypertension, 109
Hypochlorous acid (HOCl), 8, 20, 49
myeloperoxidase, 47-48
Hypoxanthine, 7

I

IB6 antibody, 73, 74
Ibuprofen, 81
Ignarro LJ, 109
Iloprost, 97
Immunoglobulins, 38, 40-42
  A (IgA), 10
Immunoglobulin superfamily, 38, 40-42
Inauen W, 49
Inositol trisphosphate (IP_3), 65
Integrin, 16, 37-38, 40-41, 72
Intercellular adhesion molecule-1 (ICAM-1), 16-17, 37, 38, 40, 41, 72
  no-reflow, 21
  -2, 37, 38
Interleukin-1 (IL-1), 94, 98
Interleukin-8 (IL-8), 97
Intravital microscopy, 35, 68
Iron (Fe), 7
  chelation, 12, 126
  delocalization, 125-126
  equilibria in cultured cells, 128-129
  in vivo sources of oxidant prone, 126-128
  NO, 106
Ischemia-reperfusion (I/R)
  complement blockade, 87, 89-90
  iron (Fe), *127f*
    intact animal, 131-133, *133f*
    isolated organs, 129-131
    therapeutic targeting, 134, 138-139
      deferoxamine (DFO), 134-136
        polymer bound, 136-138
      L1, 136
      N-methyl hexanoylhydroxamic acid, 136
  lung injury, 143, 144
    complement, 144
    neutrophil sequestration, 144-146
  nitric oxide, 110-115
  platelet-activating factor (PAF), 72-74
  remote injury, 17

J

Janoff A, 132
Johnston Jr RB, 45

K

K^+, 20
Kadsurenone, 68
Keaney Jr JF, 106
Kempner MS, 94
Klausner JM, 144
Korthuis RJ, 34, 130
Ku DD, 110

L

L1 (1,2-dimethyl-3-hydroxy-4-pyridone), 136
Lactoferrin, 45, 148, *150f*
Langendorff preparation, 129, 136
Lefer DJ, 110, 111, 129
Lehr HA, 35, 51, 138
Leukocyte adhesion, 34-35, *36t*
  no-reflow phenomenon, 20-22, *22f*
  rolling, 34, 37
Leukocyte depletion (LeuDep), *15f*, 34
Leukocyte-endothelial cell recognition, 72-74, *72t*
Leukocyte integrins (LeuCAMs), 37
Leukotrienes, 72, 146
  A_4 (LTA_4), 146
  B_4 (LTB_4), 145, 146
    clinical investigations, 148-149
    endothelium, 146
    neutrophil, 147-148
    thromboxane generation, 146-147
  chemotaxis, *150f*
  cysteinyl, 16
  D_4, 146
  F_4, 146
  synthesis, 146
Lewis RA, 145
LFA-1 (lymphocyte function-associated antigen-1), 37
Lindsay TF, 144
Lipopolysaccharide (LPS), 144
Liposomes, 10

M

Mac-1, 37, 38, 72, 80, 82
a_2-Macroglobulin, 48, 49
Majno G, 50
Major histocompatibility complex (MHC), 98
Mannitol, 149
Maruyama M, 137
Margination, 35
Marletta MA, 103
Mazur A, 131, 132
McCord JM, 3
Membrane cofactor protein, 81
Membrane permeability, 20
Metalloproteinases, *50f*
N-Methyl hexanoylhydroxamic acid, 136
Microfilaments
  cytoskeleton
    neutrophil sequestration, 17-18, *18f*
Mitchell HH, 103
Monocytes, 67
Mousa SA, 137
Mullane KM, 34
Multiple organ system failure (MOSF), 144

Myeloperoxidase (MPO), 8, 45, 83, *85f*
hypochlorous acid (HOCl), 47-48
Myocardial stunning, 130, 135, 137
Myogenic theory, 2
Myoglobin, 125

N

NAD+ (nicotinamide-adenine dinucleotide), 7
NADPH (nicotinamide-adenine dinucleotide phosphate), 8
NADPH (nicotinamide-adenine dinucleotide phosphate) oxidase, 8
reactive oxygen metabolites (ROM), 42-47
Na^+-H^+ exchange, 21
Neutrophil, 2, 3, 13, 15-18
adhesion, 34-35, *39f*
infiltration, 33-34
mechanisms of injury, 42, *42t*
no-reflow phenomenon, 21
priming, 66, 70-72, 149
rolling, 72
sequestration
in lung, 144-146
in skeletal muscle, 84, *85f*
LTB_4, 147-148
Nitric oxide (NO), 7, 19, 41, 103-105
biosynthesis, 105, *105f*
mechanisms of action, 105-106
reactive oxygen metabolite (ROM) scavenger, 111, *115f*
Nitric oxide synthase (NOS), 105,*107t*
constitutive (cNOS), 106-108
inducible (iNOS), 106-108
vasodilation, 108-110
Nitrogen dioxide (NO_2), 114
Nitroglycerine, 108
Nitroprusside, 108
L-Nitro^G-monomethyl-arginine (L-NMMA), 105, 109, 112
No-reflow phenomenon, 2, 20-21, 49-52

O

Ognibene FP, 145
OKY-046, 68
Opsonins, 78
Oxygen, 3
-derived free radicals, 6

P

Paller MS, 128, 133
Palmer RMJ, 109
Parks GN, 3
Peroxidation
membrane lipids, 19, 43-44, *43f*, 111, 114
Peroxynitrite, 106, 113, *113f*, 114, *115f*
Phalloidin, 18
Phosphatidic acid (PA), 64
Phospholipase A_2 (PLA_2), 63, 68
Phospholipase C (PLC), 64-65
Phospholipase D (PLD), 64
postischemic leukocyte infiltration, 19
Platelets, 66
nitric oxide (NO), 109
Platelet-activating factor (PAF), 19, 38, 61-62, *62f*
biosynthesis, 62-64, *63f*
microcirculatory effects, 67-68, 70-72, *68f*, *69f*, *70f*
leukocyte adhesion, 72-74
signal transduction, 64-65, *66f*
Platelet-derived growth factor (PDGF), 66
Platelet factor 4 (PF4), 66
Polyethylene glycol (PEG) polymer, 10
Polymorphonuclear leukocyte (PMN). See also Neutrophil.
Platelet-activating factor (PAF), 66-67
Postcapillary venules, 16
Postischemic leukocyte infiltration, 18-19
Properdin, 80
Prostcyclin, 109
Proteases, 20
Protein kinase C (PKC), 64, 65

R

Ramos CL, 7
Reactive (oxygen) metabolites (ROM), 6-7
postischemic leukocyte infiltration, 18-19
sources, 7
NADPH (nicotinamide-adenine dinucleotide phosphate), 8
xanthine oxidase (XO), 7-8
Reactive (oxygen) species (ROS). See Reactive (oxygen) metabolites (ROM).
Reif DW, 127
Reperfusion injury, 6, *9f*
Respiratory burst, 3, 44
Rhabdomyolysis, 7
Rinaldo JE, 145
Rheumatoid arthritis, 114
Robinson E, 133, 134, 137
Romson JL, 33

S

Sato N, 98
Secretory leukoproteinase inhibitor, 48, 49
Selectin, 35, *36ft*
E-, 17, 37, 40, 72
L-, 37, 72
P-, 17, 21, 37, 38
Sironi M, 98
Skalak TC, 51

Slow reacting substance of anaphylaxis (SRS-A), 146
Slungaard A, 98
Smith SM, 135
S-nitrosothiols, 106, 108
Snyder SH, 104
Stallone RJ, 144
Staurosporine, 65
Stolpen AH, 97
Strock PE, 50
Stuehr DJ, 103
Sulfhydryls, 106
Superoxide anion, 6
  nitric oxide (NO), 106, 110, *112f*, 114, *115f*
Superoxide dismutase (SOD), 6, 10, *11f*, 12, *12f*, 43
  iron (Fe), 127
  nitric oxide (NO), 110, 112
  postischemic leukocyte infiltration, 18-19
Synergism, 72

T

Tannenbaum SR, 103
Tetrahydrobiopterin, 107
Thiobarbituric acid reactive substances (Tbars), 138
Thomas L, 1, 3
Thrombi, 21
Thromboxane
  A_2 (TxA_2), 67, 68, *151f*
  LTB_4, 146
  synthase, 68
Transferrin, 125, 126
Transition metals, 6, 44, 126
Trauma patients
  eicosanoid levels, 148
Tsao PS, 129, 138
Tumor necrosis factor (TNF), 97, 98
Tyrosine kinase (TK), 65

V

van der Kraaij AMM, 136
Vinculin, 67
Vitamin E (a-tocopherol), 43
Voogd A, 129, 138

W

Walden DL, 34
Weibel-Palade bodies, 37, 38
Weinberg PF, 144
Weiss SJ, 48
Weselcouch EO, 41
Wiggers' graded hemorrhage model, 131
Williams DM, 127

X

Xanthine, 7, 127
Xanthine dehydrogenase, 111, 135
Xanthine oxidase (XO), 3, 7-8, 13, 111, 114
  iron (Fe), 127, 132, 133, 135

Y

Yoshida N, 40
Yoshioka T, 137

Z

Zawadski JV, 104
Zimmerman ANE, 2
Zimmerman GA, 147
Zsebo KM, 98
Zymogen-activated plasma (ZAP), 46
Zymosan, 83